MIAMI WINTER SYMPOSIA

1. *W.J. Whelan and J. Schultz, editors:* HOMOLOGIES IN ENZYMES AND METABOLIC PATHWAYS *and* METABOLIC ALTERATIONS IN CANCER, 1970
2. *D. W. Ribbons, J. F. Woessner, Jr., and J. Schultz, editors:* NUCLEIC ACID-PROTEIN INTERACTIONS *and* NUCLEIC ACID SYNTHESIS IN VIRAL INFECTION, 1971
3. *J. F. Woessner, Jr. and F. Huijing, editors:* THE MOLECULAR BASIS OF BIOLOGICAL TRANSPORT, 1972
4. *J. Schultz and B. F. Cameron, editors:* THE MOLECULAR BASIS OF ELECTRON TRANSPORT, 1972
5. *F. Huijing and E. Y. C. Lee, editors:* PROTEIN PHOSPHORYLATION IN CONTROL MECHANISMS, 1973
6. *J. Schultz and H. G. Gratzner, editors:* THE ROLE OF CYCLIC NUCLEOTIDES IN CARCINOGENESIS, 1973
7. *E. Y. C. Lee and E. E. Smith, editors:* BIOLOGY AND CHEMISTRY OF EUCARYOTIC CELL SURFACES, 1974
8. *J. Schultz and R. Block, editors:* MEMBRANE TRANSFORMATION IN NEOPLASIA, 1974
9. *E. E. Smith and D. W. Ribbons, editors:* MOLECULAR APPROACHES TO IMMUNOLOGY, 1975
10. *J. Schultz and R. C. Leif, editors:* CRITICAL FACTORS IN CANCER IMMUNOLOGY, 1975
11. *D. W. Ribbons and K. Brew, editors:* PROTEOLYSIS AND PHYSIOLOGICAL REGULATION, 1976
12. *J. Schultz and F. Ahmad, editors:* CANCER ENZYMOLOGY, 1976
13. *W. A. Scott and R. Werner, editors:* MOLECULAR CLONING OF RECOMBINANT DNA, 1977
14. *J. Schultz and Z. Brada, editors:* GENETIC MANIPULATION AS IT AFFECTS THE CANCER PROBLEM

MIAMI WINTER SYMPOSIA — VOLUME 14

GENETIC MANIPULATION AS IT AFFECTS THE CANCER PROBLEM

edited by

J. Schultz

Z. Brada

PAPANICOLAOU CANCER RESEARCH INSTITUTE
MIAMI, FLORIDA

Proceedings of the Miami Winter Symposia, January, 1977
Sponsored by The Papanicolaou Cancer Research Institute
Miami, Florida

Academic Press, Inc. New York San Francisco London 1977
A Subsidiary of Harcourt Brace Jovanovich, Publishers

Academic Press Rapid Manuscript Reproduction

ACADEMIC PRESS, INC.
111 Fifth Avenue, New York, New York 10003

United Kingdom Edition published by
ACADEMIC PRESS, INC. (LONDON) LTD.
24/28 Oval Road, London NW1

Library of Congress Cataloging in Publication Data

Main entry under title:

Genetic manipulation as it affects the cancer problem.

(Miami winter symposia ; v. 14)
Includes bibliographical references.
1. Oncogenic viruses–Congresses. 2. Genetic engineering–Congresses. 3. Cancer–Genetic aspects–Congresses. I. Schultz, Julius, Date II. Brada, Zbynek. III. Papanicolaou Cancer Research Institute. IV. Series. [DNLM: 1. Genetic intervention –Congresses. 2. Recombination, Genetic–Congresses. 3. DNA replication–Congresses. 4. Neoplasms–Congresses. W3 MI202 v. 14 1977 / QZ206 G328 1977]
RC268.57.G46 616.01'94 77-10005
ISBN 0-12-632755-6
PRINTED IN THE UNITED STATES OF AMERICA

CONTENTS

SPEAKERS, CHAIRMEN, AND DISCUSSANTS ix
PREFACE xiii

Construction and Biological Activities of Deletion Mutants of Simian Virus 40 1
D. Nathans, C.-J. Lai, W.A. Scott, W.W. Brockman, and S.P. Adler
DISCUSSION: J. Davison, P. Berg, C.A. Thomas, and P. Duesberg

Site-Directed Mutagenesis as a Tool in Genetics 11
C. Weissman, T. Taniguchi, E. Domingo, D. Sabo, and R.A. Flavell
DISCUSSION: R. Kavenoff

Simian Virus 40 as a Cloning Vehicle in Mammalian Cells 37
D.H. Hamer
DISCUSSION: U. Littauer, P. Berg, and P. Duesberg

Use of Defective SV40 Replicons for the Propagation of Prokaryotic DNA in Mammalian Cells 49
G.C. Fareed and P. Upcroft
DISCUSSION: P. Berg, D. Nathans, and J. Sambrook

HR-T Mutants of Polyoma Virus 73
T.L. Benjamin
DISCUSSION: L. Miller, P. Sarin, and J. Davison

The Molecular Basis of Transformation by Simian Virus 40 87
R.G. Martin, M. Persico-DiLauro, C.A.F. Edwards, and A. Oppenheim
DISCUSSION: P. Duesberg, A. Koch, D. Billen, and S. O'Brien

Interaction of SV40 T Antigen With Selected Regions of SV40 DNA . . . 103
D.M. Livingston, D.G. Tenen, D. Jessel, L.L. Haines, V. Woodward, A.P. Modest, A. Maxam, and J. Hudson
DISCUSSION: A. Bollon, C. Weissmann, D. Hamer, K. Sakaguchi, and R. Kavenoff

Studies of Adenovirus and SV40 Genes Required for *in Vitro* Transformation 121
A.J. van der Eb, J.H. Lupker, J. Maat, P. Abrahams, H. Jochemsen, A. Houweling, W. Fiers, C. Mulder, H. van Ormondt, and A. de Waard

SV40-Adenovirus 2 Hybrid Viruses. 139
J. Sambrook

The Genetic Map of Rous Sarcoma Virus 161
P.H. Duesberg, L.-H. Wang, P. Mellon, W.S. Mason, and P.K. Vogt
DISCUSSION: C. Weissman and S. O'Brien

Recombination Events between Simian Virus 40 and the Host Genome . 181
E. Winocour, M. Oren, S. Lavi, T. Vogel, and Y. Gluzman
DISCUSSION: P. Duesberg, J. Hassell, J. Schultz, and M. Singer

The Importance of Transcription Unit Definition 197
J.E. Darnell, J. Weber, S. Goldberg, W. Jelinek, N. Fraser, and P. Sehgal
DISCUSSION: N. Chiu, D. Roufa, P. Duesberg, J. Hassell, and C. Wei

PANEL DISCUSSION 217
Chairmen: D. Stetten, Jr. and J. Schultz
Panelists: A. Bayev, P. Berg, B. Davis, M.F. Singer, R.L. Sinsheimer, and J. Tooze

DISCUSSION: Decker, R. Curtiss,
V. Sgaramella, and L. Jacobs

FREE COMMUNICATIONS

Transfection of Permissive Monkey Cells with Restriction Endonuclease-Derived Fragments of SV40 DNA.
I. Transformation and Superinfection Properties of Recipient Cells. 261
R.C. Moyer, M.P. Moyer, and H. Hurtado

Transfection of Permissive Monkey Cells with Restriction Endonuclease-Derived Fragments of SV40 DNA.
II. Rescue of SV40 Virions by Cell Fusion and Transfection. 263
M. Moyer, R. Moyer, M. Gerodetti, and G. Lipotich

In Vitro Recombinant DNA Technology For Mapping Animal Virus Mutations 265
L.K. Miller

Nuclear Translational Units of Adenovirus-Infected HeLa Cells 267
N.K. Chatterjee, H.W. Dickerman, and T.A. Beach

Transcription and Coupled Transcription-Translation of Cloned DNAs in *Xenopus* Oocytes 269
J.E. Mertz, J.B. Gurdon, and E.M. De Robertis

A Safer Model System for Studying the Effects of Recombining Animal Virus DNA 271
L.K. Miller

A Role for Cell Surface Topography for Directing the Synthesis of Anti-Glycosyl Antibodies 273
J.H. Pazur and K.L. Dreher

SPEAKERS, CHAIRMEN, AND DISCUSSANTS

F. Ahmad (Session Co-Chairman), Papanicolaou Cancer Research Institute, Miami, Florida

A. Bayev (Panel Member), USSR Academy of Sciences, Moscow, USSR

P. Berg (Panel Member), Stanford University Medical Center, Stanford, California

T.L. Benjamin, Harvard Medical School, Boston, Massachusetts

D. Billen, Oak Ridge National Laboratory, Oak Ridge, Tennessee

A. Bollon, University of Texas, Dallas, Texas

R.E. Block (Session Co-Chairman), Papanicolaou Cancer Research Institute, Miami, Florida

Z. Brada (Session Co-Chairman), Papanicolaou Cancer Research Institute, Miami, Florida

N. Chiu, National Institutes of Health, Bethesda, Maryland

P. Curtiss, University of Alabama, Birmingham, Alabama

J.E. Darnell, The Rockefeller University, New York, New York

B. Davis (Panel Member), Harvard University, Cambridge, Massachusetts

J. Davison, Institute of Cellular Pathology, Brussels, Belgium

Decker, Chemical Abstracts Service, Columbus, Ohio

P.H. Duesberg (Session Chairman), University of California, Berkeley, California

G.C. Fareed, Harvard Medical Center, Boston, Massachusetts

D.H. Hamer, Harvard Medical School, Boston, Massachusetts

J. Hassell, Cold Spring Harbor Laboratory, Cold Spring Harbor, New York

L. Jacobs, National Institutes of Health, Bethesda, Maryland

R. Kavenoff, University of California, San Diego, California

A. Koch, Indiana University, Bloomington, Indiana

U.Z. Littauer, The Children's Hospital of Philadelphia, Philadelphia, Pennsylvania

D.M. Livingston, Harvard Medical School, Boston, Massachusetts

R.G. Martin, National Institutes of Health, Bethesda, Maryland

L.K. Miller, University of Idaho, Moscow, Idaho

D. Nathans (Session Chairman), Johns Hopkins University School of Medicine, Baltimore, Maryland

S. O'Brien, National Institutes of Health, Bethesda, Maryland

D. Roufa, Kansas State University, Manhattan, Kansas

K. Sakaguchi, Mitsubishi-Kasei Institute for Life Sciences, Tokyo, Japan

J. Sambrook, Cold Spring Harbor Laboratory, Cold Spring Harbor, New York

P. Sarin, National Institutes of Health, Bethesda, Maryland

J. Schultz (Panel Co-Chairman), Papanicolaou Cancer Research Institute, Miami, Florida

V. Sgaramella, Laboratorio di Genetica, Pavia, Italy

M.F. Singer (Panel Member), National Institutes of Health, Bethesda, Maryland

R.L. Sinsheimer (Panel Member), California Institute of Technology, Pasadena, California

D. Stetten, Jr. (Panel Chairman), National Institutes of Health, Bethesda, Maryland

C.A. Thomas (Session Chairman), Harvard Medical School, Boston, Massachusetts

J. Tooze (Panel Member), European Microbiology Organization, Heidelberg, West Germany

A.J. van der Eb, Sylvius Laboratoria der Rijksuniversiteit, Leiden, The Netherlands

C.M. Wei, National Institutes of Health, Bethesda, Maryland

C. Weissmann, Universität Zurich, Zurich, Switzerland

E. Winocour, The Weizmann Institute of Science, Rehovot, Israel

PREFACE

As we go to press we are very pleased with the fact that this volume (and the companion volume assmbled by the Biochemistry Department of the University of Miami Medical School) reports the first meeting of the principal architects in the new and exciting field of genetic manipulation. Included are the reports of those participants selected by the Papanicolaou Cancer Research Institute to speak at the ninth Annual Miami Winter Symposia. They are devoted to laboratory findings and represent the present state of the art.

The topic for these symposia was conceived about 18 months prior to the annual meeting organized in collaboration with the Department of Biochemistry of the University of Miami Medical School. In those months following the Asilomar Conference the well-publicized series of conferences took place that culminated in the NIH Guidelines. The controversies at Cambridge, Princeton, Ann Arbor, and California attracted a great deal of attention from the press, television, and nationally circulated periodicals. During this same period the co-directors of these symposia appeared on television to inform their respective institutions and the people of the Miami area of the significance of research on recombinant DNA.

To further illuminate the issues, the Papanicolaou Cancer Research Institute allocated its final symposia session to a panel discussion, which is reported verbatum as part of this volume. The panel members are acknowledged experts who have not only been very active in the field of recombinant DNA, but have also played important roles in establishing the current Guidelines. They have appeared before Congress and on national television and have been quoted by the press. Their comments here provide a reasonably sound analysis of the issue and include both pro and con viewpoints regarding the need for continuing recombinant DNA research. Their presentations before an audience of knowledgeable scientists afforded them an opportunity to gain varying reactions. We hope that such discussions will reduce the hysteria created by those opportunists seeking attention through unfounded statements designed to panic the uninformed. We wish to acknowledge the suggestions of Professor William Whelan in selecting participants for the panel discussion.

The international cross section of the speakers and the audience participating at these symposia provided a great forum for critical discussion. Both volumes should offer the scientific community a record unavailable elsewhere, at this time, of the pioneering effort that is building the firm foundation for still another magnificent epoch in the history of the natural sciences.

We would like to acknowledge the support of the Boehringer Mannheim Corporation, Hoffman-LaRoche Inc., Eli Lilly and Company, Abbott Laboratories, SmithKline Corporation, and the Upjohn Company. Finally, we offer our deep appreciation to our typists, Kathi Bishop, Anne Johnson, and Gloria Teixeira, who worked under the supervision of Mrs. Ginny Salisbury.

J. Schultz
Z. Brada

CONSTRUCTION AND BIOLOGICAL ACTIVITIES OF DELETION MUTANTS OF SIMIAN VIRUS 40

D. NATHANS, C.-J. LAI, W.A. SCOTT, W.W. BROCKMAN
and S.P. ADLER
Department of Microbiology
Johns Hopkins University School of Medicine
Baltimore, Maryland 21205 U.S.A.

Abstract: Deletion mutants of Simian Virus 40 (SV40) have been cloned from serially passed virus stocks or constructed by enzymatic manipulation of SV40 DNA. Some of the mutants produce detectable short polypeptides derived from SV40-coded proteins. Mutants in the late region have been used to map the junction between the D and B/C genes and to define SV40 functions which require only the early or A gene. Mutants lacking portions of both late genes but with an intact early genomic segment were able to induce T antigen in infected cells, replicate their DNA in the absence of helper virus, stimulate thymidine incorporation in cellular DNA, and transform mouse and hamster cells. Cells transformed by late deletion mutants were shown to contain the mutant genome by a cell fusion-complementation rescue procedure. Deletion mutants lacking substantial portions of the early genomic region, including those segments where tsA mutants map, lacked all of the above activities.

INTRODUCTION

The genome of SV40 is a circular DNA duplex of about 5,000 nucleotide pairs. The genetic organization of this molecule has been reviewed recently (1), and is summarized in the form of a physical map in Fig. 1. As seen in the figure, there is a single origin for SV40 DNA replication, one early gene (A gene), which codes for T antigen, and two late genes (B/C and D), which code for the major virion protein (VP1) and minor virion protein (VP2/3), respectively. Also shown in the figure are the map positions of early and late SV40 messenger RNA and the map positions of temperature-sensitive mutants, which have helped to delineate the positions of structural genes. We have been using cloned defective deletion mutants constructed by enzymatic manipulation or isolation from serially passed virus to further localize SV40 genes and to assign the various biological activities of the virus to specific regions of the genome. In this communication we summarize our experiments on the properties of mutants containing deletions of either the early or late region of the SV40 genome.

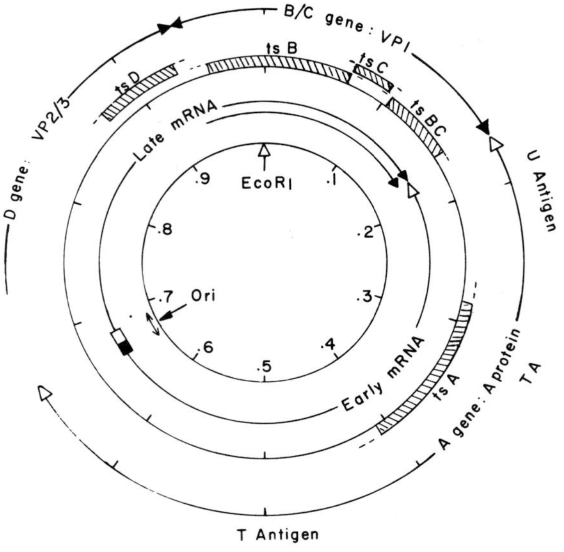

Fig. 1. Physical map of the SV40 genome (from ref. 1).

RESULTS AND DISCUSSION

Deletion mutants of SV40 have been isolated by plaque formation in the presence of a complementing helper virus (2). From serially passed virus a number of mutants (called evolutionary variants or ev) have been isolated which are missing segments of one or more viral genes and often have duplications around the origin of SV40 DNA replication. For construction of simple deletion mutants, SV40 DNA has been linearized by specific or nonspecific cleavage with endonuclease, or by the single strand specific enzyme S1 after random nicking of the circular form I DNA molecule, as illustrated in Fig. 2. Generation of deletion mutants by this procedure is dependent on a poorly understood cyclization reaction in infected cells whereby a linear molecule is circularized with loss of nucleotides at each end (3, 4, 5). This cell-mediated activity has allowed the construction of deletion mutants from all portions of the SV40 genome.

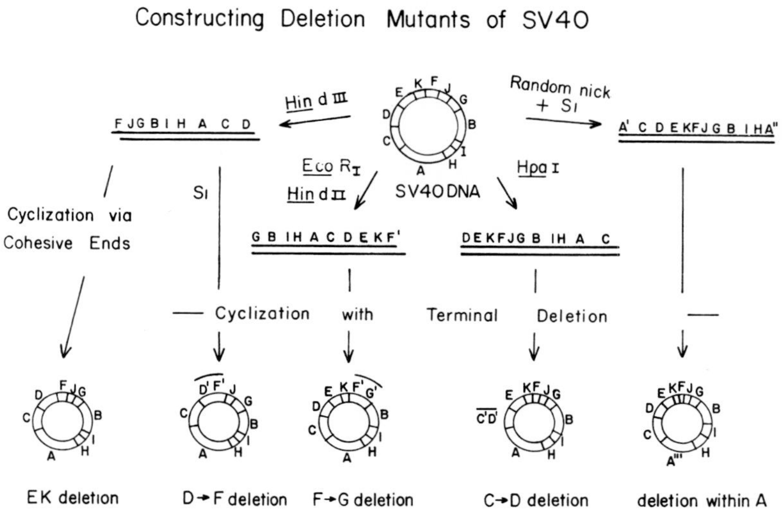

Fig. 2. Construction of deletion mutants of SV40 by enzymatic cleavage of viral DNA and cell-mediated cyclization of the linear molecule.

The map positions of several late and several early deletion mutations are shown in Fig. 3. Several of the mutants shown were tested for their ability to produce short SV40 proteins in infected cells. The results with two such mutants are illustrated in Fig. 4. dl-1010 which has a deletion of about half of the region where tsB, C and BC mutants map, produces a short VP1 polypeptide as illustrated in the figure (6). dl-1010, a mutant in which that segment of the early region where tsA mutants map has been deleted was shown by Rundell et al. (7) to produce a short polypeptide with T antigen activity, as illustrated in Fig. 4. Whereas the dl-1010 peptide may have arisen by in-phase reading beyond the deletion, in the case of dl-1010 the new 33,000 dalton T antigen may have arisen by premature termination beyond the deletion. The new T antigen could also be a specific cleavage product of some larger polypeptide. In any case, these results help establish the structural gene products of SV40.

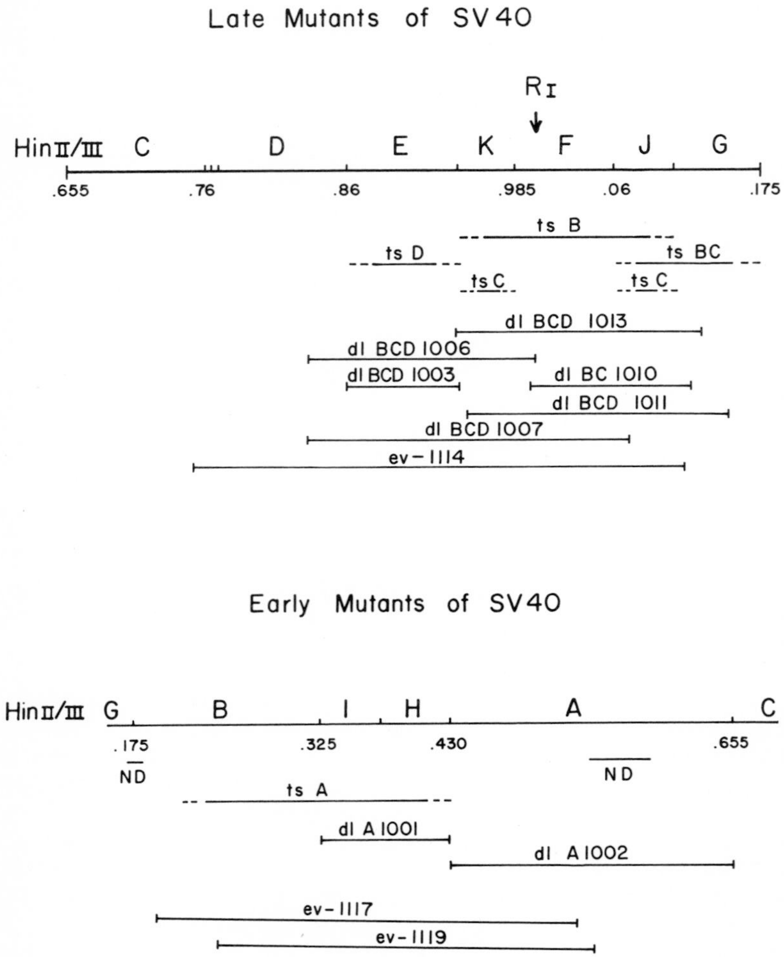

Fig. 3. Map positons of SV40 deletion mutations compared to the HindII/III cleavage map. dl are constructed mutants; ev are evolutionary variants. Letters preceding mutant number indicate the complementation group(s) of the mutant. ND refers to the map position of nondefective deletion mutants described by Shenk et al. (14).

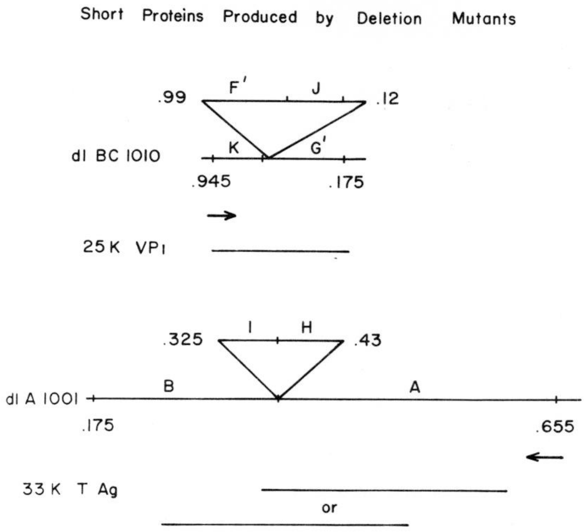

Fig. 4. Diagram of deletion mutants that produced short proteins in infected cells. The map positions of each deletion are given by the triangles. Letters K, F, J, G and B, I, H, A refer to HindII/III fragments of SV40 DNA. In each case the new protein was shown to share tryptic peptides with the reference SV40 protein (VPI for dl-1010 and T antigen for dl-1010).

Complementation tests between deletion mutants and temperature-sensitive mutants have helped define the limits of structural genes, and in one case have pointed to the presence of a regulatory signal in the DNA preceding the structural gene (6). This approach is illustrated in Fig. 3 and Fig. 5, in which the position of deletion mutants in and around the start of the B/C gene are shown in relation by Van de Voorde et al. (8). As seen in Fig. 5, mutant dl-1010 which begins at approximately 0.995 map units within Hin-F complements tsD mutants but not B/C mutants, whereas dl-1011 which begins at about 0.945 map units within Hin-K complements neither D nor B/C mutants. Coming from the other side of the B/C gene it is seen that mutant dl-1003 which ends at 0.945 map units failed to complement B as well as D mutants, even though the deletion does not extend into the coding sequence of the B/C gene. In the case of dl-1003 the precise limits of the deletion are known, since this mutant is an excisional deletion mutant constructed by excision of Hin-E fragment by HindIII restruction endonuclease

(3). These results suggest first, that the D gene may actually overlap with the B/C gene and secondly, that within the Hin-E segment preceding the B/C gene proper there is a signal required for expression of the B/C gene. Since it is known that the messenger RNA for the D gene product (VP2 and 3) and the B/C gene product (VP1) is initially linked (9), one explanation for both of the above findings is that a nucleotide sequence signal is present in the region immediately preceding the initiation codon for VP1 that is involved in processing and thus activating late messenger RNA.

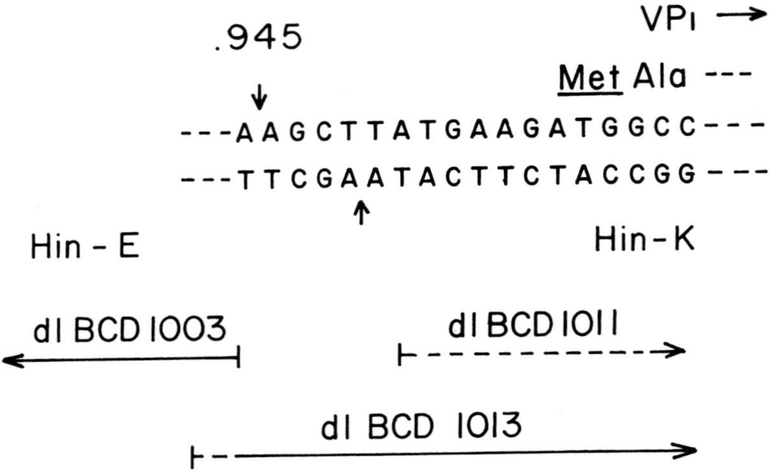

Fig. 5. Map positions and complementation groups of mutants near the start of the B/C (VP1) gene in relation to the nucleotide sequence of this region. The precise leftward boundary of dl-BCD 1011 is not known, but does not extend into the HindIII sequence A A G C T T at the left.

To test the biological activities of large deletion mutants (dl-1007, ev-1114, ev-1117, and ev-1119), these mutants were separated from helper virus in the mutant stock by cesium chloride equilibrium centrifugation. The resulting purified deletion mutant virions were then assayed by complementation with appropriate temperature-sensitive mutants and used to infect permissive or nonpermissive cells at a predetermined multiplicity (10). Several SV40 functions were measured: 1) viral DNA replication independent of helper virus, 2) stimulation of thymidine incorporation into cellular DNA, 3) T antigen formation, and 4) cell transformation. In the case of cells transformed by mutant viruses the viral genome was rescued by fusing transformed cells with permissive cells and superinfecting with a complementing temperature-sensitive mutant of SV40. The results are summarized in Table 1. As shown in the table, the mutants lacking segments of the early region of SV40, namely, ev-1117 and ev-1119, were unable to replicate in the absence of helper virus, stimulate thymidine incorporation into cellular DNA, form T antigen in infected permissive cells, or transform nonpermissive mouse cells or semipermissive hamster cells. In contrast, mutants which lacked segments of both of the two late genes, namely, dl-1007 and ev-1114 were similar to wild type SV40 in all of the above activities. Our general conclusion from these results is that the early regions of SV40 plus continguous DNA segments is sufficient for all of the above viral functions. A similar conclusion follows from previous work with segments of SV40 DNA by van der Eb and his colleagues (11) or with temperature-sensitive mutants of SV40 (12,13).

TABLE 1

Mutant	Genome segment missing (map units)	T antigen	Viral DNA replication	Stimulation of 3T3 DNA synthesis	Transformation		Rescue of mutant from transformed cells	
					3T3	CHL	3T3	CHL
ev-1114	0.75-0.11	+	+	+	+	+	+	+
dl-1007	0.83-0.07	+	+	+	+	+	+	+
ev-1117	0.18-0.53	-	-	-	-	-		
ev-1119	0.24-0.54	-	-	-	-	-		

More recently we have constructed a series of smaller early deletion mutants of SV40 by the procedures illustrated in Fig. 2. Preliminary results with these mutants indicate that the A gene, defined by complementation tests with tsA mutants extends throughout the early region as defined by transcription mapping. These mutants are now being tested for their ability to produce altered T antigen polypeptides and for biological activities in an attempt to further dissect the multiple early functions of SV40.

REFERENCES

(1) T.J. Kelly, Jr. and D. Nathans. Adv. in Virus Res. (1976) in press.

(2) W.W. Brockman and D. Nathans. Proc. Nat. Acad. Sci. U.S. 71 (1974) 942.

(3) C.-J. Lai and D. Nathans. J. Mol. Biol. 89 (1974) 179.

(4) N. Murray and K. Murray. Nature 251 (1974) 476.

(5) J. Carbon, T.E. Shenk and P. Berg. Proc. Nat. Acad. Sci. U.S. 72 (1975) 1392.

(6) C.-J. Lai and D. Nathans. Virology 75 (1976) 335.

(7) K. Rundell, J.K. Collins, P. Tegtmeyer, H. Ozer, C.-J. Lai and D. Nathans. J. Virol. (1977) in press.

(8) A. Van de Voorde, R. Contreras, R. Rogiers and W. Fiers. Cell 9 (1976) 117.

(9) G. Khoury, B.J. Carter, F.J. Ferdinand, P.M. Howley, M. Brown and M.A. Martin. J. Virol 17 (1976) 832.

(10) W.A. Scott, W.W. Brockman and D. Nathans. Virology 75 (1976) 319.

(11) P.J. Abrahams, C. Mulder, A. Van de Voorde, S.O. Waarner and A.J. van der Eb. J. Virol. 16 (1975) 818.

(12) P. Tegtmeyer. J. Virol. 15 (1975) 613.

(13) R.G. Martin and J.Y. Chou. J. Virol. 15 (1975) 599.

(14) T.E. Shenk, J. Carbon and P. Berg. J. Virol. 18 (1976) 644.

The research summarized in this report was supported by grants from the U.S.P.H.S. National Cancer Institute (CA 16519), the American Cancer Society (VC-132A), and the Whitehall Foundation.

Discussion

J. Davison, Institute of Cellular Pathology, Brussels: Dr. Nathans, you showed that late deletions of SV40 are able to transform non-permissive cells. Have you done that same experiment with permissive cells?

D. Nathans, Johns Hopkins University School of Medicine: We have not.

J. Davison: I would not expect it to because I do not believe that you can have an initiation function and the origin of replication both present in an integrated provirus in permissive cells and still have a viable cell.

D. Nathans: Well, I can tell you that the mutant that is lacking almost the entire late region causes normal CPE in permissive cells and of course it replicates.

P. Berg, Stanford University: In rescuing the EV1114 integrated genomes from transformed cells, do you find any host sequences in the viral DNA recovered from the plaques?

D. Nathans: Dr. Berg, we have not looked for that, but Gary Getner and Thomas Kelly have been pursuing that approach for about the last year and a half, starting with wild type transformed cells, transformed at a low multiplicity of infection. They have collected a very large number of abnormal genomes, but they have not determined as yet whether they contain cell DNA. It looks like a possibly promising approach to finding joints between cell DNA and SV40 DNA after misexcision.

P. Berg: May I ask just one more question? Are any of the mutants having deletions in the early region viable?

D. Nathans: Yes, we found viable deletions, they were not shown on that map. Those were all defectives. We have just put the viable ones aside into a separate collection and have not done anything with them.

C.A. Thomas, Harvard Medical School: May I ask a question? Do you want to say anything about the cellular DNA which you noted was non-reiterated DNA? Does that mean that it is not the same sequence from virus molecule to virus molecule or does it belong to the non-reiterated class of the cellular DNA?

D. Nathans: Non-reiterated operationally means that it is not highly reiterated in a cot test with cellular DNA. This is a cloned variant, so within the variant it is one sequence.

P. Duesberg, University of California, Berkeley: You have described cells which had been infected with two viral mutants, tsA and a deletion mutant. I did not see in your table any wild type recombinants showing up. I would expect to see wild type recombinants among the progeny.

D. Nathans: I did not dwell on this point, but in the table I showed what comes out of that transformed clone in the complementation fusion type of rescue procedure, one gets a whole series of things. We got recombinants, and revertants of the helper virus because we had to put in a large excess of helper virus.

SITE-DIRECTED MUTAGENESIS AS A TOOL IN GENETICS

C. WEISSMANN, T. TANIGUCHI, E. DOMINGO[a], D. SABO[b] and R.A. FLAVELL[a]
Institut für Molekularbiologie I
Universität Zürich
8093 Zürich, Switzerland

Abstract: We describe a method for the generation of nucleotide transitions at predetermined positions of a viral RNA. Qβ RNA with an A⟶G replacement in position 40 from the 3' terminus, an extracistronic region, was infectious and gave rise to viable mutant phage. This mutant was outgrown by wild type phage in a few generations under competitive conditions, possibly because of impaired RNA replication. Qβ RNA with a G⟶A substitution in position 16 from the 3' terminus was non-infectious, despite the fact that its RNA was efficiently replicated <u>in vitro</u>. Qβ RNA with a G⟶A mutation in the third nucleotide of the coat cistron, which converted the initiator triplet AUG into AUA, was no longer bound to ribosomes under conditions of protein initiation, however a double mutant, in which the third and fourth G residues were substituted by A was bound with about 30% efficiency relative to wild type RNA. The application of site-directed mutagenesis to DNA is outlined.

---

[a]Present address of E.D.: Instituto de Biología del Desarrollo, Madrid-6, Spain; of R.A.F.: Jan Swammerdam Institute, University of Amsterdam, Amsterdam, The Netherlands.

[b]Died February 11th, 1976.

The advent of hybrid DNA technology portends a breakthrough in our understanding of the structure and function of the eukaryotic genome. It will undoubtedly soon be possible to integrate any gene of interest, with its neighboring regions, into an appropriate replicon *in vitro*, clone and amplify it and reintroduce it into eukaryotic cells. This constitutes the basis for what may be designated as "new genetics", an approach wherein DNA regions are modified at predetermined positions *in vitro* and the effects of these interventions are scored *in vivo* or *in vitro*, in contradistinction to classic genetics, where deviant phenotypes are first isolated, and the lesion giving rise to them is identified subsequently.

The modifications of the DNA may be gross, such as deletions or insertions of DNA segments introduced by the techniques of *in vitro* recombination, or point mutations or mutations limited to a short, localized region, generated by site-directed mutagenesis. In this paper we will summarize some experiments in site-directed mutagenesis as applied to the RNA phage Qβ and outline how the technique may be applied to hybrid plasmids.

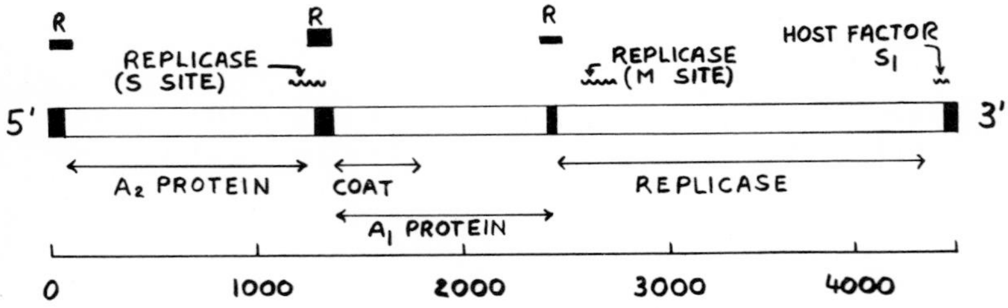

Fig. 1. Map of Qβ RNA. Non-translated areas are black; the cistrons are indicated by double-headed arrows. The ribosome binding sites are marked R, binding sites for Qβ replicase, host factor and protein S_1 (a component of Qβ replicase) are indicated by wavy lines. Based on refs. 1, 13, 27.

I. Some facts about phage Qβ and its replication.*

Phage Qβ, a small spherical virus, contains an RNA molecule of about 4500 nucleotides which serve both as genome and messenger RNA. As shown in Fig. 1, Qβ RNA consists of 3 translatable and 4 non-translatable (extracistronic) segments. While the regions immediately preceding the cistrons are involved in the initiation and regulation of protein synthesis, the function of the longer untranslatable segments at the ends of the genome is not known. It has been speculated that the precise conservation of these sequences is essential for the viability of RNA phages (4).

After penetrating its host, the viral RNA first serves as messenger RNA. As shown *in vitro*, ribosomes initially bind exclusively at the initiation site of the coat cistron; binding at the replicase cistron occurs only after coat translation has begun, and the A_2 cistron is probably only translated off nascent RNA strands. RNA replication begins after Qβ replicase has been assembled from the phage-coded subunit and 3 host-specific components.

Purified Qβ replicase, in conjunction with a host factor, replicates Qβ RNA *in vitro*, yielding infectious progeny RNA in large excess over the input template. Qβ replicase shows little affinity for the 3' end of Qβ RNA, where RNA synthesis begins, but binds tightly to 2 internal sites of Qβ RNA, one of which (S site) partly overlaps the coat cistron ribosome binding site (cf. Fig. 1). Presumably this interaction places the 3' terminus of the RNA into the initiation site of the polymerase. The product of the first step of synthesis, a single- stranded Qβ minus strand, is non-infectious but serves as an excellent template for the synthesis of infectious Qβ RNA. The stability *in vitro* of the replication complexes at all stages allows manipulations to be carried out with little loss of enzymatic activity.

---

*) For original references, see the reviews by Weissmann et al. (1), Weissmann (2) and Kamen (3).

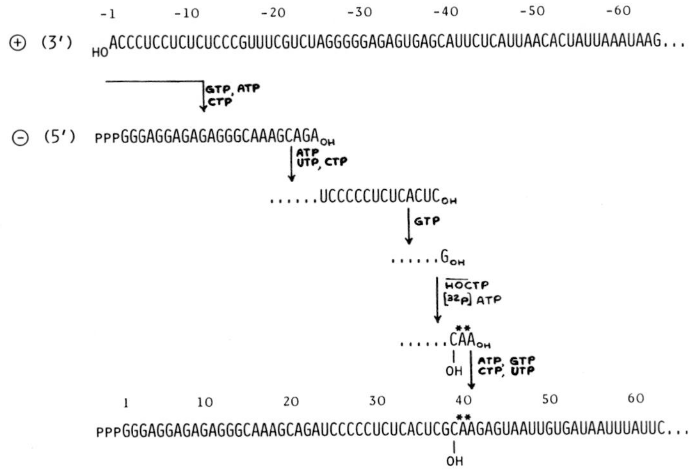

Fig. 2. Scheme for the stepwise synthesis of Qβ minus strands with introduction of HOCMP in position 39 from the 5' terminus.

II. Site-directed mutations in Qβ RNA.

As pointed out above, little is known about the functions of the non-translated regions in RNA phages. We have therefore generated Qβ RNAs with point mutations in the 3' terminal extracistronic region and studied their viability *in vivo*. In addition, we have prepared Qβ RNAs with mutations located in the coat cistron initiator region and tested their ribosome binding capacity.

a) Qβ RNA with an A⟶G substitution in position 40 from the terminus*.

Stepwise synthesis of Qβ minus strands with N4-hydroxyCMP in position 39.

As shown schematically in Fig. 2, incubation of Qβ replicase with host factor, Qβ RNA as template, and with GTP, ATP and CTP as only substrates led to the synthesis of a 23 nucleotide-long minus strand segment extending

*) The details are given in ref. 5.

to the position where UTP would be required for further elongation. After removal of the substrates by Sephadex chromatography the nascent minus strand was elongated up to position 37 with ATP, CTP and UTP. The complex was again purified and the 38th residue, GMP, added. The substrate was removed and the complex incubated with N^4-hydroxyCTP (in lieu of UTP) and [α-^{32}P] ATP, to add the sequence pHOC$\overset{*}{p}$A$\overset{*}{p}$A$_{OH}$ which could be later detected by nucleotide analysis. The complex was reisolated, the minus strands were completed with the 4 standard triphosphates and purified free of plus strands. Digestion of a sample with RNAase T_1 yielded HOC$\overset{*}{p}$A$\overset{*}{p}$ApGp as main radioactive oligonucleotide (70% of the total radioactivity) showing that the nucleotide analog had efficiently replaced UMP in position 39.

Synthesis of Qβ plus strands using minus strands as template.

The purified, substituted minus strands were used as template for the synthesis of one round of ^{32}P-labeled plus strands. The product was purified, digested with RNAase T_1, and the oligonucleotides were fractionated by two-dimensional polyacrylamide gel electrophoresis. A comparison of the resulting fingerprint with that of wild type RNA (cf. ref. 5) showed that a new large oligonucleotide, designated T1*, had appeared, while the amount of oligonucleotide T1 was diminished. The ratio of T1* to T1 was 1 : 3. Fig. 3 shows the positions of these two oligonucleotides in the fingerprint. T1 is derived from positions -63 to -38 at the 3' end of wild type Qβ RNA (1, 6)(cf. Fig. 2) and has the sequence

    -60   -50
A-A-U-A-A-A-U-U-A-U-C-A-C-A-A-U-U-A-C-U-C-U-
-40   -60
U-A-C-Gp. The structure of T1* was A-A-U-A-A-
                    -50   -40
A-U-U-A-U-C-A-C-A-A-U-U-A-C-U-C-U-U-Gp. Therefore, an A⟶G transition at position -40 of

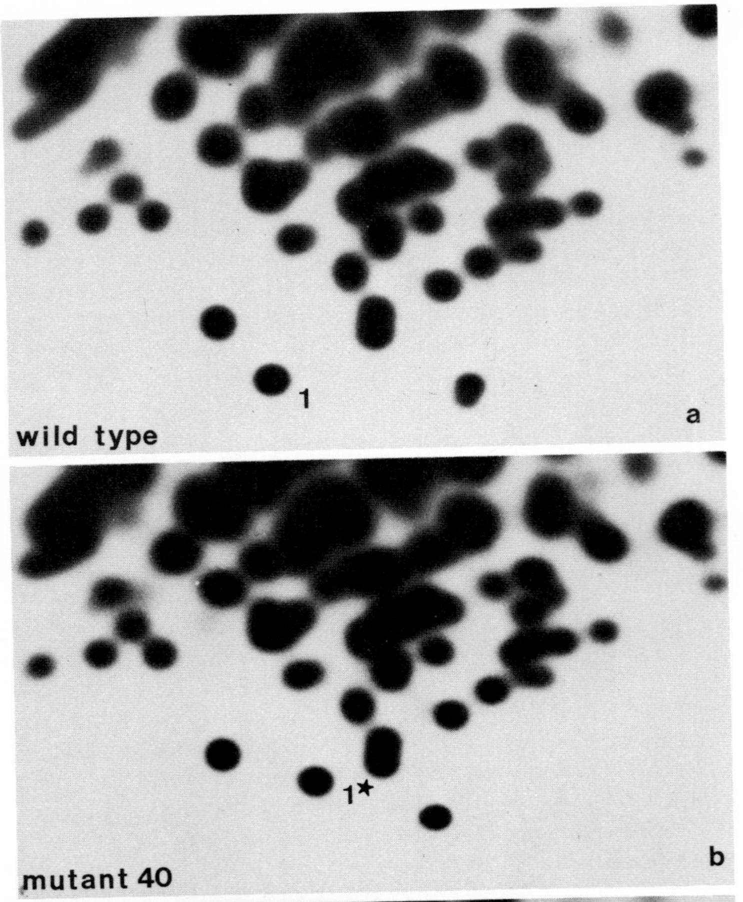
1
a
wild type
1*
b
mutant 40
1*
wild type
+
mutant 40
1
c

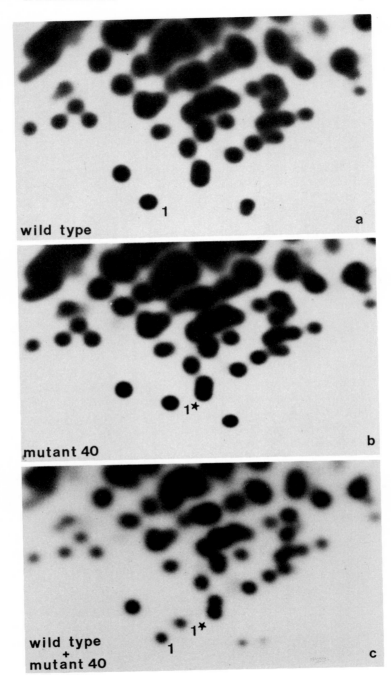

the wild type sequence accounts for the appearance of the new oligonucleotide. Since the synthesis of the minus strands of RNA phages starts at the penultimate nucleotide of the plus strand (7-9), the 40th nucleotide from the 3' end of the plus strand is complementary to the 39th position from the 5' end of the minus strand, i.e., the position into which N^4-hydroxyCMP had been introduced (cf. Fig. 2). Thus, the analog directed the incorporation of either GMP or AMP into the complementary position of the plus strand.

Infectivity of plus strands synthesized with N^4-hydroxyCMP-substituted minus strands as template.

As first shown by Feix et al. (10) Qβ minus strands, which are inherently non-infectious, can serve as template for the synthesis of infectious plus strands *in vitro*. Minus strands substituted with N^4-hydroxyCMP in position 39 were incubated with Qβ replicase and GTP and ATP as the only substrates, then polyethylene sulfonate was added (to inhibit initiation (11) and thus prevent further replication) and the chains were elongated by the addition of CTP and [α-^{32}P] UTP. No infectivity was detectable in the spheroplast assay prior to the addition of the missing triphosphates, proving the absence of biologically active plus strands in the template (Table 1). After the reaction a total of 9 ng of plus strands had been formed per 120 ng of minus strands used as template; the specific

Fig. 3. Two-dimensional polyacrylamide gel electrophoresis of the T_1 oligonucleotides of uniformly ^{32}P-labeled wild type RNA and mutant (A_{-40}→G) RNA prepared from cloned phage. (a) Wild type RNA, (b) mutant (A_{-40}⟶G) RNA, (c) a mixture of wild type and mutant (A_{-40}⟶G) RNA (from ref. 5).

TABLE 1

Infectivity of Qβ RNA plus strands generated by a single round of in vitro synthesis using $\overline{HO}$CMP-substituted or wild type minus strands as template.

Template	Substrates	[^{32}P] RNA synthesized (ng/2 μl)	Infectious units	Specific Infectivity[a] (pfu/ng of product)
			(pfu/5 μl)	
(a) Wild type minus strands	GTP, ATP	-	0; 1	-
	GTP, ATP, CTP, [α-^{32}P] UTP	1.9	116; 110	25
			(pfu/4 μl)	
(b) Substituted minus strands	GTP, ATP	-	0; 0	-
	GTP, ATP CTP, [α-^{32}P] UTP	1.5	100; 104	33

Wild type (0.3 μg) or $\overline{HO}$CMP-substituted (0.2 μg) minus strands, 2 units of Qβ replicase, 0.8 mM each of GTP and ATP, 80 mM Tris-HCl (pH 7.5), 12 mM MgCl, 1 mM EDTA (final volume 20 μl) were incubated for 5 min at 37°C. Polyethylene sulfonate was added to prevent further initiation. Aliquots were taken for the determination of infectivity, and CTP and [α-^{32}P] UTP (1.2 x 10^6 cpm/nmole) were added to allow completion of plus strands. Acid-insoluble radioactivity and infectivity were determined in duplicate (details in ref. 5).

[a] The specific infectivity of Qβ RNA extracted from virions was 27 pfu/ng.

infectivity of the product was similar to that of plus strands synthesized on wild type minus strands or plus strands from virions.

<u>Isolation of phage carrying an A⟶G transition in position -40.</u>

The ^{32}P-labeled RNA of 18 phage clones generated by the infection of spheroplasts with the RNA synthesized on substituted minus strands was examined by T_1 fingerprinting. Four of the 18 preparations showed T1*, diagnostic for the mutant $A_{-40}\rightarrow G$, and no significant amounts of T1; 14 RNAs gave rise to T1 but not to T1*. This proportion of mutant phage, 22%, reflects rather accurately the proportion of mutant RNA, 25%, determined chemically in the preparation used for transfection.

<u>Competitive growth *in vivo* between mutant $A_{-40}\rightarrow G$ and wild type phage.</u>

To test if wild type phage had a selective advantage over mutant 40 in vivo, E.coli Q13 was infected with a 1 : 1 mixture of cloned mutant and wild type phage at an overall multiplicity of infection of about 20 phage per cell, to ensure intracellular competition. The resulting lysate was then used to infect a fresh culture at the same multiplicity. A total of 10 such cycles of infection was carried out. Cultures infected with mutant 40 alone were propagated in parallel. ^{32}P-labeled RNA was prepared from several of the intermediate lysates and analyzed to determine the proportion of mutant. As shown in Fig. 4, after four cycles of infection the mutant content was only about 2% and no mutant was detected after ten cycles. Propagation of carefully recloned mutant phage $A_{-40}\rightarrow G$ in the absence of added wild type reproducibly resulted in the appearance of an increasing proportion of wild type phage after a few cycles, showing that revertants arose at a substantial rate and outgrew the mutant. From the data of Fig. 4 we have estimated a reversion rate of 10^{-4} per doubling and a growth rate of 0.25 of the mutant relative to

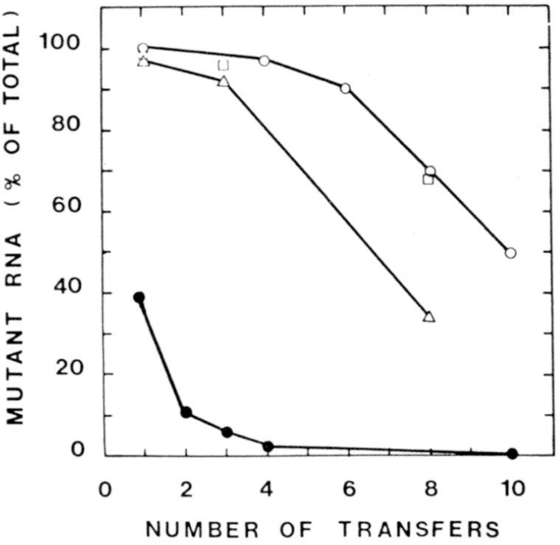

Fig. 4. Competitive growth of mutant Qβ (A_{-40} ⟶ G) and wild type Qβ phage <u>in vivo</u>. Wild type and mutant phage were isolated from single plaques. <u>E.coli</u> were infected with either mutant phage or a 1 : 1 mixture of mutant and wild type phage at a multiplicity of 15-25. The resulting lysate was used to infect a fresh bacterial culture at the same multiplicity; this procedure was reported for a total of 10 transfers. The content of mutant was determined by T_1 fingerprint analysis of ^{32}P-labeled RNA. Experiment 1, o—o, infection with mutant A_{-40}→G (clone 1); Experiment 2, □——□, infection with mutant (A_{-40} ⟶ G) (phage repurified by end-point dilution of clone 1); Experiment 3, △—△, infection with mutant (A_{-40} ⟶ G) (end-point repurified clone 2); Experiment 4, ●——●, infection with a mixture of a wild type clone (m.o.i. ≃ 10) and mutant (A_{-40} ⟶ G) (clone 1) (m.o.i. ≃ 10). (From ref. 5).

wild type under competitive conditions (12).

A serial transfer experiment under non-competitive conditions was not carried out, however we found no significant difference in the burst size of <u>E.coli</u> singly infected with wild type or mutant phage.

Competitive replication in vitro of wild type and mutant ($A_{-40} \rightarrow G$) Qβ RNA.

One possible explanation for the reduced propagation rate of mutant ($A_{-40} \longrightarrow G$) phage is that its RNA is replicated less effectively by the replicase under competitive conditions.

In order to test the relative replication rates of wild type and mutant RNA, a mixture of the two was used as a template for Qβ replicase in a serial transfer synthesis, in which each transfer corresponded to a doubling of the input RNA. T_1 fingerprint analyses of the products at different stages showed that the mutant RNA content decreased from 52% after the first transfer to 30% after the 18th (5). Since one cycle of infection in vivo involves the equivalent of about ten to fifteen doublings of the input RNA, the disadvantage of the mutant phage in vivo could be roughly accounted for by the slower replication of its RNA, as measured in vitro.

It is of interest that the nucleotide in position -40 is part of a sequence (-63 to -38) which binds both host factor I and S_1 protein (13). Host factor I is required by Qβ replicase for initiation on plus strands (14); its interaction with the 3' terminal region of Qβ RNA, as well as with a further sequence (13) located in the middle of the RNA, may be required to bring the RNA into the proper conformation for initiation. S_1 is a ribosomal protein (15-17), which, after infection, is recruited as the α-subunit of Qβ replicase (18, 19). Goelz and Steitz (personal communication) have recently found that the mutant oligonucleotide T1* is bound less efficiently by protein S_1 than its wild type counterpart, T1, suggesting that the reduced efficiency of RNA replication could be due to weaker binding of replicase and/or host factor to the mutated binding site.

b) <u>Qβ RNA with a G⟶A substitution in position 16 from the 3' terminus*.</u>

<u>Generation of the mutant (G_{-16}⟶A) Qβ RNA.</u>

In this instance minus strand synthesis was initiated with GTP and ATP only, leading to elongation up to position 14 inclusive, at which point CTP was required for further elongation (Fig. 5). After removal of the unused substrates the sequence $p\overline{HOC}\overset{*}{p}A\overset{*}{p}A\overset{*}{p}A_{OH}$ was added by incubating with $\overline{HOC}$TP and [α-^{32}P] ATP. A large excess of CTP and ATP was added along with GTP and UTP and incubation was continued to allow completion of the minus strand. Analysis of the 30 S minus strands showed that 78% of the nucleotide in position 15 was $\overline{HOC}$MP, the rest being CMP, presumably derived from traces of CTP in the $\overline{HOC}$TP preparation. The newly synthesized minus strand was purified from the plus strand template and used

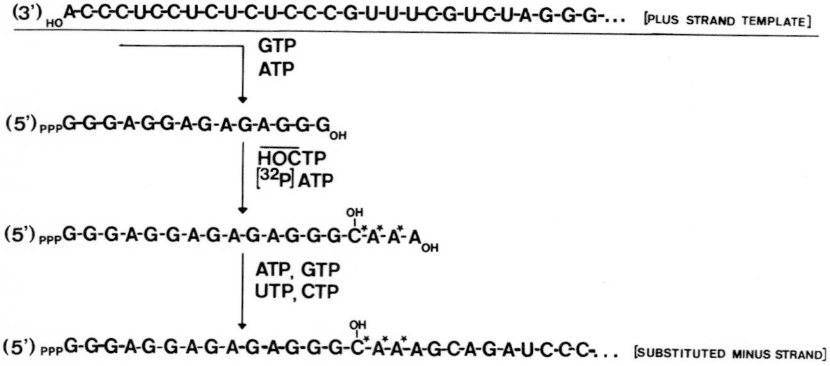

Fig. 5. Scheme for the stepwise synthesis of Qβ minus strands with introduction of $\overline{HOC}$MP in position 15 from the 5' terminus. (From ref. 21).

*) The details are given in refs. 20 and 21.

to direct one round of plus strand synthesis by Qβ replicase. T_1 fingerprints of the product showed the appearance of a new oligonucleotide T18', with the sequence C-U-U-U-A-C-C-C-U-C-U-C-U-C-C-U-C-C-C-A_{OH}, and reduction of the amount of oligonucleotide T18, C-C-C-U-C-U-C-U-C-C-U-C-C-C-A_{OH}, the terminal segment of Qβ RNA. It is evident from the sequence shown in Fig. 5 that a G⟶A transition in position -16 would cause this change in the T_1 RNAase cleavage pattern. The ratio of T18' to T18 was 1 : 2, showing that about 33% of the plus strands carried the mutation.

Properties of mutant (G_{-16}⟶A) Qβ RNA*.

In order to determine whether Qβ RNA with the extracistronic G_{-16}⟶A mutation was infectious, spheroplasts were infected with the first generation of plus strands synthesized on minus strands with HOCMP-substitution in position -15. As in the previous experiment (Table 1), the incubation mixtures contained no infectious units before incubation with all four triphosphates; the product resulting after elongation had only half the specific infectivity of wild type plus strands (Table 2). Seventy-eight clones derived from spheroplasts infected with RNA from a one-round synthesis and 42 clones from infections with RNA from multiple rounds of synthesis were analyzed by T_1 fingerprinting: all were wild type. Since the RNA preparations used for transfection contained 30% and 60%, respectively, of the mutant (G_{-16}⟶A) RNA and no mutant clones were found, plaque formation efficiency of mutant relative to wild type RNA is less than about 0.03. The failure to find mutant clones cannot be ascribed to the manipulations to which the RNA had been subjected, since the specific infectivity of wild type RNA synthesized *in vitro* on the substituted minus strand was about

*) Details of this experiment are given in Sabo et al. (submitted for publication).

TABLE 2

Infectivity of plus strands generated by a single round of synthesis using minus strands substituted with HOCMP in position 15 as template[a].

Minus strand template	Conditions of synthesis	Incorporation of [^{32}P] UMP (pmoles/2 μl)	Specific infectivity[b] (pfu/ng product)
A. wild type	initiation only	N.D.	0
	initiation and elongation	5.7	25
B. HOC-substituted	initiation only	N.D.	0
	initiation and elongation	6.3	9

a) The experiment was carried out in principle as described in Table 1, using 0.4 μg of wild type and 1 μg of HOC-substituted minus strands, respectively, in a final reaction mixture of 20 μl. The specific activity of the [^{32}P] UTP was 8×10^5 cpm/nmol.

b) The specific infectivity of natural Qβ RNA was 22 pfu/ng.

equal to that of natural Qβ RNA. Moreover, the results obtained with the mutant ($A_{-40} \rightarrow G$) RNA described above showed that the procedures used can in fact yield viable mutants.

In order to determine the efficiency of replication of the mutant ($G_{-16} \rightarrow A$) RNA by Qβ replicase, a serial transfer experiment was initiated with a mixture of mutant and wild type RNA and the ratio of the two was determined by T_1 fingerprinting. Contrary to the experiment with the mutant ($A_{-40} \rightarrow G$) RNA, the proportion of mutant RNA increased with successive transfers, and reached a value of about 80% after extensive replication (21). Thus the deleterious effect of the ($G_{-16} \rightarrow A$) transition on the infectivity of the RNA does not appear to be due to impaired RNA replication.

c) <u>Qβ RNA with mutations in the coat cistron initiator region*.</u>

Under initiation conditions, <u>E.coli</u> ribosomes bind to Qβ RNA almost exclusively at the coat cistron initiation site. To determine to what extent the AUG triplet is required for 70 S complex formation, we prepared Qβ RNA with G→A transitions of the 3rd and 4th nucleotides of the coat cistron, i.e., with modifications in the third position of the A-U-G codon and the following nucleotide (cf. Fig. 6).

<u>Synthesis of Qβ minus strands with $\overline{HO}$CMP substitutions in the positions complementary to the 3rd and 4th nucleotide of the coat cistron.</u>

In order to carry out stepwise synthesis in the required region we synchronized minus strand synthesis at a ribosome attached to the coat initiation site (22). As outlined in Fig. 7, the 70 S Qβ RNA ribosome complex was

*) Details to be given in Taniguchi & Weissmann (manuscript in preparation).

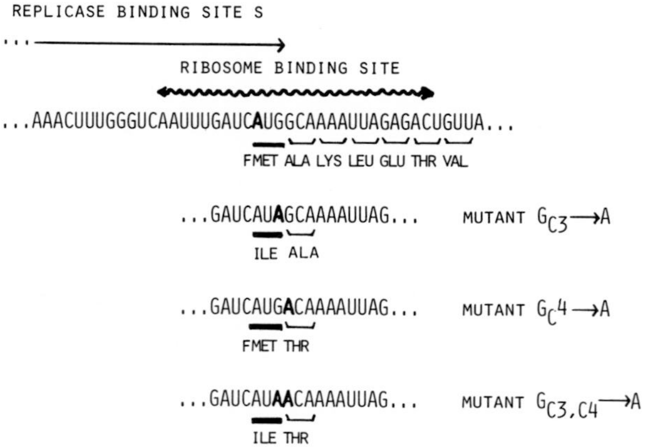

Fig. 6. Nucleotide sequence around the ribosome binding site of the coat cistron of wild type Qβ RNA, and mutants generated by site-directed mutagenesis at C_3 and C_4. The nucleotide sequence was established by Hindley & Staples (25) and Weber et al. (23).

used as template for Qβ replicase. Synthesis proceeded up to the position corresponding to the 16th nucleotide of the coat cistron (22). The ribosome was then dislodged by treatment with EDTA and stepwise synthesis was carried out as shown in Fig. 7b, leading to insertion of $\overline{\mathrm{HOC}}$MP into the positions complementary to the 4th and 3rd nucleotides of the coat cistron. Analysis of the appropriately labeled minus strands showed that in the region of interest 30% of the RNA had the sequence ...$\overline{\mathrm{HOC}}$-$\overline{\mathrm{HOC}}$-A-U-G..., 15% had ...C-$\overline{\mathrm{HOC}}$-A-U-G.., 25% ...$\overline{\mathrm{HOC}}$-C-A-U-G... and 32% ...C-C-A-U-G...; the presence of CMP is due to traces of CTP which were difficult to remove in this experiment and competed effectively with $\overline{\mathrm{HOC}}$TP.

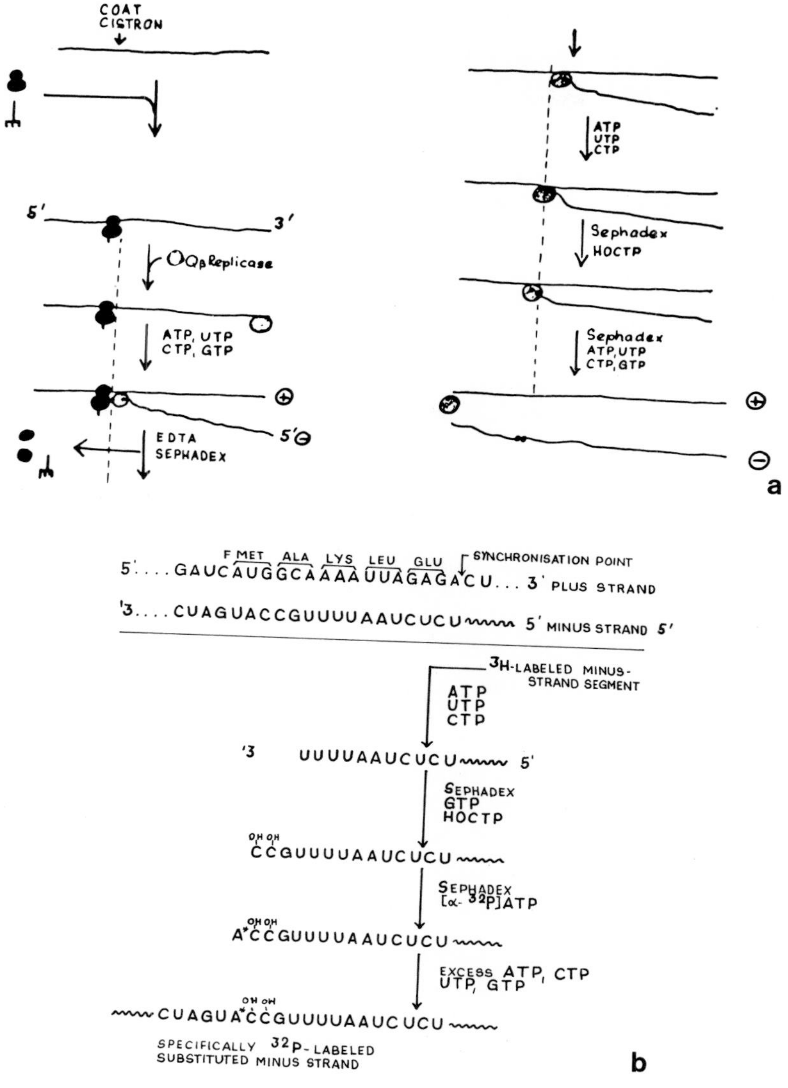

Fig. 7. Scheme for the synchronization of Qβ minus strand synthesis at the coat ribosome binding site and stepwise synthesis with introduction of HOCMP at positions complementary to the 3rd and 4th nucleotide of the coat cistron. See text for explanation. The black spheres represent ribosomes, the open circles Qβ replicase, the fork-like symbol, fmet tRNA.

Analysis and characterization of plus strands synthesized with the substituted minus strands as template.

Qβ plus strands were synthesized on substituted minus strands and replicated *in vitro*. In order to analyze the region comprising the coat cistron initiation site, we took advantage of the fact that when Qβ replicase is bound to Qβ RNA at 0.1 M NaCl in the absence of Mg^{++} and the complex digested with RNAase T_1, a 100-nucleotide long fragment, the S-site (Figs. 1 and 6) remains bound to the enzyme (23). This fragment extends up to the A-U-G at the beginning of the coat cistron, and with appropriate nearest neighbor analysis of ^{32}P AMP-labeled plus strand the sequence at positions C_3 and C_4 could be determined. The ratio of wild type : mutant ($G_{C_3} \longrightarrow G$) : mutant ($G_{C_4} \longrightarrow A$) : mutant ($G_{C_3,C_4} \longrightarrow A$) was found to be 1 : 1.8 : 1.6 : 4.5 (average of 3 experiments). The presence of the double mutant ($G_{C_3,C_4} \longrightarrow A$) RNA could be demonstrated directly by comparing the T_1 fingerprints of wild type and mutant RNA. As shown in Fig. 8, the mutant RNA preparation gave a new large T_1 oligonucleotide, Q1, for which the structure A-U-C-A-U-A-A-C-A-A-A-A-U-U-A-G was deduced. Inspection of the sequence around the beginning of the

C_1

coat cistron, ...U-G-A-U-C-A-U-G-G-C-A-A-A-

C_{10}

A-U-U-A-G... shows that G $\longrightarrow$ A transitions at positions C_3 and C_4, would eliminate two T_1 cleavage sites and account for the appearance of a new T_1 oligonucleotide (underlined) with the sequence of Q_1 (cf. Fig. 6). The RNAs with a single transition would not be expected to yield large new oligonucleotides.

We have not yet determined whether any of the mutant RNAs are infectious.

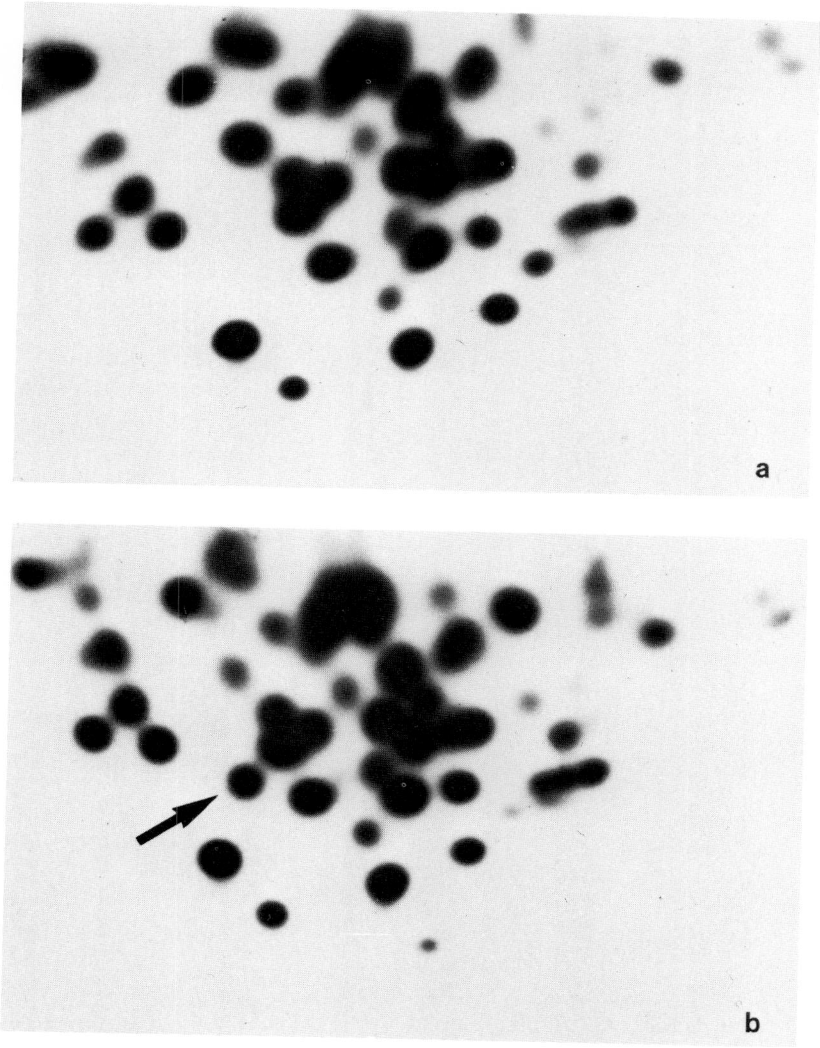

Fig. 8. Two-dimensional polyacrylamide gel electrophoresis of the T_1 oligonucleotides of (a) ^{32}P CMP-labeled wild type RNA and (b) a preparation of ^{32}P CMP-labeled Qβ RNA with G⟶A substitutions in positions C_3 and C_4. The arrow indicates the new large T_1 oligonucleotide Q1, A-U-C-A-U-A-A-C-A-A-A-A-U-U-A-G.

The ribosome binding capacity of Qβ RNA with G⟶A transitions in positions C_3 and/or C_4.

A preparation of ^{32}P-labeled Qβ RNA consisting of the species described above was bound to ribosomes under initiation conditions, the 70 S complex was treated with RNAase A and the RNA fragment retained by the ribosome was isolated (24, 25) and analyzed with regard to the C_3/C_4 region. All RNA fragments recovered were derived from the coat initiation region.

The composition of the ribosome-bound fragments with regard to this region (average of 3 experiments) was wild type : mutant (G_{C3}⟶A) : mutant (G_{C4}⟶A) : mutant ($G_{C3,C4}$⟶A) = 1 : < 0.1 : 4.4 : 1.5. Taking into consideration the original composition of the RNA preparation, we estimate the relative binding efficiencies to be 1 : < 0.1 : 2.8 : 0.33.

Both mutant RNAs lacking the A-U-G triplet have a reduced ribosome binding capacity, but the double mutant RNA is still bound with considerable efficiency. This suggests that the $tRNA_F^{met}$-AUG interaction contributes substantially to the stabilization of the 70 S ribosome complex, but that other interactions, perhaps the ones postulated by Shine & Dalgarno (26), may suffice for the formation of a less stable 70 S complex at the correct site.

It is striking that ribosomes are bound more efficiently to mutant (G_{C4}⟶A) RNA than to wild type RNA,and to mutant ($G_{C3,C4}$⟶A) more efficiently than to mutant (G_{C3}⟶A) RNA. Perhaps the nucleotides flanking the codon and the anticodon contribute to the stability of the interaction with fmet-tRNA. As shown below,the mutant RNAs (G_{C4}⟶A) and ($G_{C3,C4}$⟶A) can form an additional A-U base pair as compared to wild type RNA and mutant (G_{C3}⟶A) RNA, respectively.

```
3'            5'          3'            5'
 \C-U-A-C-U/               \C-U-A-C-U/
    |  |  ||                  |  |  || |
 -C-A-U-G-G-               -C-A-U-G-A-

5'            3'          5'            3'
```

wild type $\qquad$ $(G_{C4} \rightarrow A)$ mutant

```
3'            5'          3'            5'
 \C-U-A-C-U/               \C-U-A-C-U/
    |  |                      |  |     |
 -C-A-U-A-G-               -C-A-U-A-A-

5'            3'          5'            3'
```

$(G_{C3} \rightarrow A)$ mutant $\qquad$ $(G_{C3,C4} \rightarrow A)$ mutant

III. Some perspectives for site-directed mutagenesis.

The work described above has demonstrated the feasibility of generating point mutations in Qβ RNA using the principle of substrate-controlled RNA synthesis and introduction of a mutagenic nucleotide, HOCMP, in place of either CMP or UMP in a pre-determined position. Other nucleotide analogs, substituting for purine nucleotides, should also become available. This procedure yields a mixture of wild type and mutant RNA which can be used as such for some studies; if the mutant phage RNA is viable it can be cloned *in vivo* and obtained in a pure form. We have also developed a biochemical procedure, based on selective elongation, for purifying mutant RNA (Sabo et al., submitted for publication), however since this method requires extensive *in vitro* replication of the RNA, it is possible that additional undesired mutations accumulate in the preparation.

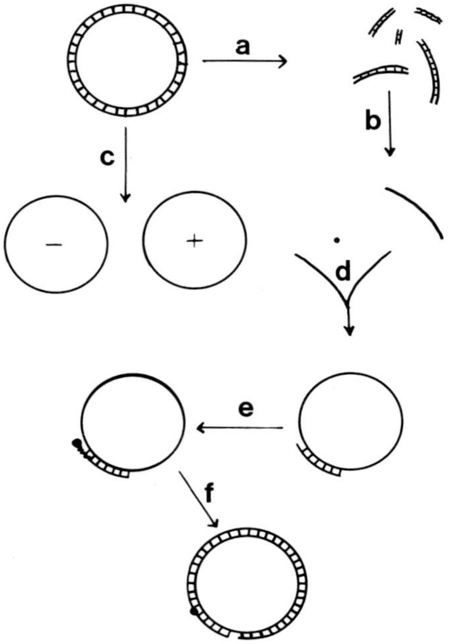

Fig. 9. Scheme for site-directed mutagenesis of a DNA plasmid. (a) The plasmid is cleaved by an appropriate restriction enzyme. (b) Restriction fragments are denatured, fractionated by agarose gel electrophoresis and the appropriate single stranded restriction fragment is isolated. (c) Plasmid is nicked, denatured and the intact circles are isolated by centrifugation through a denaturing gradient. (d) The single-stranded restriction fragment is annealed with the circular DNA, yielding a primer-template which is (e) elongated in a substrate-controlled, stepwise fashion with DNA polymerase. After insertion of the nucleotide analog in the appropriate position, standard triphosphates are used to reconstitute the double-stranded circular DNA (f). This DNA is cloned and mutant DNA is identified by sequence analysis, starting in from the appropriate restriction site.

We believe that the application of our methodology to DNA, in particular to hybrid DNA, will prove to be feasible and useful. Fig. 9 shows an approach to site-directed DNA mutagenesis currently being pursued in our laboratory by W. Müller and H. Weber. Access to the region of interest is obtained with use of specific restriction enzymes. In principle, a single-stranded, purified restriction fragment is hybridized to denatured, circular DNA and used as primer for DNA polymerase. Substrate-controlled synthesis, insertion of mutagenic nucleotide analogs at a single position or in a more extended region, and completion of the DNA chain is then carried out, in principle as described in the case of Qβ RNA. The resulting plasmids are cloned and the mutants identified by sequence analysis. It is obvious that with use of hybrid plasmids, mutations which would be deleterious or lethal in a eukaryotic cell can be generated, cloned and investigated _in vitro_ or, when appropriate vectors become available, _in vivo_ in eukaryotic cells.

Acknowledgements: We thank Dr. H. Weber for critical reading of the manuscript. This work was supported by the Schweizerische Nationalfonds (No. 3.475-0.75).

REFERENCES

(1) C. Weissmann, M.A. Billeter, H.M. Goodman, J. Hindley and H. Weber, Annu. Rev. Bioch., 42 (1973) 303-328.

(2) C. Weissmann, FEBS Lett., 40 (1974) S10-S18.

(3) R.I. Kamen, in: RNA Phages, ed. N.D. Zinder (Cold Spring Harbor Laboratory, 1975) pp. 203-234.

(4) W. Min Jou, G. Haegeman, M. Ysebaert and W. Fiers, Nature, 237 (1972) 82-88.

(5) E. Domingo, R.A. Flavell and C. Weissmann, Gene (in press).

(6) H.M. Goodman, M.A. Billeter, J. Hindley and C. Weissmann, Proc. Nat. Acad. Sci. USA, 67 (1970) 921-928.

(7) R. Kamen, Nature, 221 (1969) 321-325.

(8) U. Rensing and J.T. August, Nature, 224 (1969) 853-856.

(9) H. Weber and C. Weissmann, J. Mol. Biol., 51 (1970) 215-224.

(10) G. Feix, R. Pollet and C. Weissmann, Proc. Nat. Acad. Sci. USA, 59 (1968) 145-152.

(11) M. Kondo & C. Weissmann, Europ. J. Biochem., 24 (1972) 530-537.

(12) E. Batschelet, E. Domingo and C. Weissmann, Gene (in press).

(13) A.W. Senear and J.A. Steitz, J. Biol. Chem., 251 (1976) 1902-1912.

(14) M.T. Franze de Fernandez, W.S. Hayward and J.T. August, J. Biol. Chem., 247 (1972) 824-831.

(15) H. Inouye, Y. Pollack and J. Petre, Europ. J. Biochem., 45 (1974) 109-117.

(16) A.J. Wahba, M.J. Miller, A. Niveleau, T.A. Landers, G.G. Carmichael, K. Weber, D.A. Hawley and L.I. Slobin, J. Biol. Chem., 249 (1974) 3314-3316.

(17) J.M. Hermoso and W. Szer, Proc. Nat. Acad. Sci. USA, 71 (1974) 4708-4712.

(18) R.I. Kamen, Nature, 228 (1970) 527-533.

(19) M. Kondo, R. Gallerani and C. Weissmann, Nature, 228 (1970) 525-527.

(20) R.A. Flavell, D.L. Sabo, E.F. Bandle and C. Weissmann, J. Mol. Biol., 89 (1974) 255-272.

(21) R.A. Flavell, D.L.O. Sabo, E.F. Bandle and C. Weissmann, Proc. Nat. Acad. Sci. USA, 72 (1975) 367-371.

(22) D. Kolakofsky, M.A. Billeter, H. Weber and C. Weissmann, J. Mol. Biol. 76 (1973) 271-284.

(23) H. Weber, M. Billeter, S. Kahane, J. Hindley, A. Porter and C. Weissmann, Nature New Biol., 237 (1972) 166-170.

(24) J.A. Steitz, Nature, 224 (1969) 957-964.

(25) J. Hindley and D.H. Staples, Nature, 224 (1969) 964-967.

(26) J. Shine and L. Dalgarno, Nature, 254 (1975) 34-38.

(27) J. Hindley, D.H. Staples, M.A. Billeter and C. Weissmann, Proc. Nat. Acad. Sci. USA, 67 (1970) 1180-1187.

Discussion

R. Kavenoff, University of California, San Diego: I would like to ask you whether your mutant 16 is as efficient as wild-type QB in synthesizing full-length transcripts _in vitro_? The fingerprints you showed of mutant RNA after only a few generations and then after 20 generations show a number of changes, suggesting some sequences are not often made "late" in the reaction. Is this equally true for wild-type?

C. Weissman, University of Zurich: Yes.

R. Kavenoff: I would like to ask you about mutant 16, the second one that you described. Regarding _in vitro_ synthesized RNA, my question is whether you get the same frequency of full length transcripts from _both_ mutant and wild-type? In particular, is the mutant as efficient as wild-type in making full-length RNA?

C. Weissman: You are very observant. What you are pointing out is perfectly correct. As one replicates mutant RNA over 20 generations, one starts picking up additional changes in the RNA due to replication errors. What we really wanted to do was to make enough of this non-infectious mutant RNA so that we could then study its properties _in vitro_. Since, on amplification of the RNA, additional mutations seem to have been generated, these studies could not be carried out since we could not tell whether the original or the subsequent mutations determined the properties of the RNA preparation. Nonetheless we can deduce from the transfer experiment that mutant 16 RNA is replicated better than wild type RNA because the mutant RNA has this advantage from the first replication on.

If you replicate pure wild type RNA _in vitro_ it also starts picking up mutations and, as Spiegelman pointed out many years ago, once you relieve the phage RNA of selective pressures imposed upon it by replication _in vivo_, it is free to mutate and optimize itself to serve as a template for replication only.

SIMIAN VIRUS 40 AS A CLONING VEHICLE IN MAMMALIAN CELLS

DEAN H. HAMER

Laboratory of Molecular Genetics
National Institute of Child Health and Human Development
National Institutes of Health
Bethesda, Maryland 20014

Abstract: Simian virus 40 (SV40) genomes from which portions of the late region have been excised can be used as cloning vehicles in mammalian cells. Recombinants formed with such vehicles are propagated as virus by coinfection with conditional lethal early gene mutant helper DNA under non-permissive conditions. This system provides three useful features:

1. It is possible to select for recombinant viruses, even though it is not possible to select for the functioning of the inserted DNA fragment.

2. The foreign DNA can be transcribed, possibly due to read-through from a viral promoter.

3. The recombinants retain all of the information needed for transformation, and can be used to propagate the foreign DNA in non-permissive cells.

INTRODUCTION

Specialized transducing bacteriophage are among the most powerful tools available for analyzing bacterial genes. They can be used to enrich for particular regions of the host chromosome, move genes from one location to another, create stable diploids, and, in some cases, overproduce desired gene products (1). These phage arise spontaneously following abnormal excision of a prophage (2), and can also be constructed biochemically (3-5).

It seems reasonable to speculate that eukaryotic transducing viruses might provide equally useful in studying eukaryotic genes. How might such viruses be isolated? The genetic techniques traditionally used to construct and select for transducing phage are not available in eukaryotes. An alternative approach is to make the transducing virus *in vitro*, by enzymatically joining the desired DNA fragment to a viral DNA molecule capable of infecting and replicating in eukaryotic cells.

Recently this approach has been used to form recombinants between the well known monkey virus SV40 and sundry fragments of prokaryotic DNA (6-10). This paper describes the use of SV40 genomes from which portions of the late region have been excised as cloning vehicles in mammalian cells. The utility of such vehicles is illustrated by studies of a recombinant between SV40 and a fragment of *E. coli* DNA carrying the $tRNA^{Tyr}$ nonsense suppressor gene *su⁺III*; these experiments are described in detail elsewhere (8,9).

THE CLONING SYSTEM

As shown in Fig. 1, fragments between 700 and 2000 base pairs (bp) long can be removed from the SV40 late gene region by sequential cleavage with two restriction endonucleases such as *BamI* plus *EcoRI*, *BamI* plus *HpaII*, or *HpaII* plus *EcoRI*. Each of these enzymes has a single site in the SV40 late region, and produces fragments with cohesive

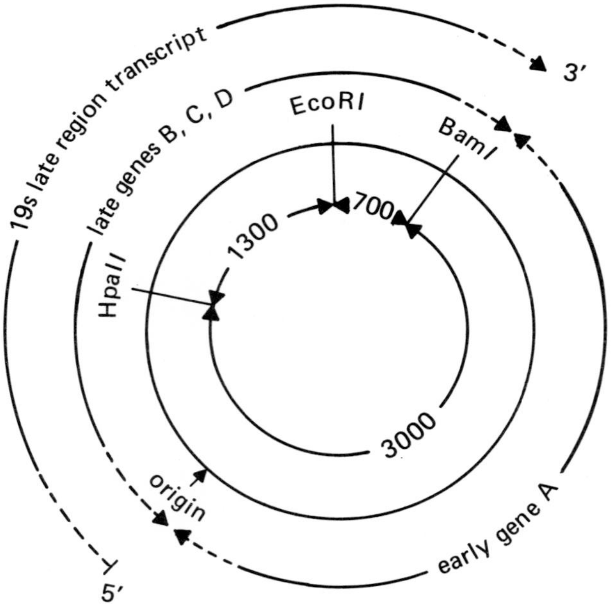

Fig. 1. A map of the SV40 genome.

termini. This generates cloning vehicles which retain the origin of viral DNA replication and the early gene A, but lack sequences necessary for the expression of at least one of the late genes B, C, and D (11-16). The foreign DNA fragment is ligated to the vehicle, and the recombinant molecules are used to infect monkey cells together with SV40 tsA as the helper. The infection is performed at 41°, which is non-permissive for the helper. Under these conditions, virus is produced only by cells infected with both the recombinant, which provides early gene function, and the helper, which provides late gene functions. This yields a mixed virus lysate which can subsequently be used to infect cells for the preparation of virus-specific DNA and RNA.

This late region deletion vehicle plus conditional early gene mutant helper system provides three especially useful features, which are described below.

SELECTION FOR RECOMBINANT VIRUSES

Since virus is produced only by cells mixedly infected with recombinant and helper DNA, it is possible to select for recombinant virus even though it is not possible to select for the functioning of the inserted DNA fragment.

The strength of this selection is shown in Fig. 2. In this experiment, an 870 bp fragment of *E. coli* DNA carrying the $tRNA^{Tyr}$ $su^{+}III$ gene was ligated to a 3700 bp fragment of SV40 DNA produced by cleavage with *Hpa*II and *Eco*RI, and circular recombinant molecules were separated from linear fragments in a dye density gradient. Monkey cells were infected, at 41°, with the recombinant molecules plus SV40 tsA as helper, or with recombinant or helper DNA alone as controls. After 12 days, the cultures were freeze-thawed, and the resulting lysates were used to infect fresh cultures of monkey cells. These were Hirt extracted (17) 2 days later, and covalently closed circular DNA was purified and analyzed by electrophoresis through agarose gels. There are no viral species present in the DNA recovered from cells infected with the recombinant alone or helper alone lysates. In contrast, two viral species are evident in the DNA recovered from cells infected with the recombinant plus helper lysate. About 70% of the viral DNA has the same length as helper (5000 bp), whereas the remainder has the length expected for the SV40-$su^{+}III$ recombinant (4570 bp). Thus, the recombinant and helper complemented one another, and the recombinant DNA was encapsidated into virus particles capable of replicating in monkey cells after the secondary infection.

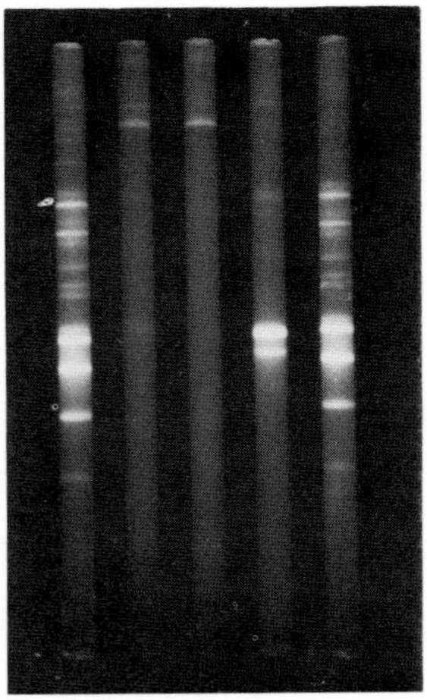

Fig. 2. Propagation of an SV40 transducing virus. An 870 bp fragment of *E. coli* DNA carrying the $tRNA^{Tyr}$ $su^{+}III$ gene and having one *EcoRI* terminus and one *HpaII* terminus was ligated to the 3700 bp SV40 vehicle produced by cleavage with *EcoRI* plus *HpaII*. Recombinant molecules were separated from linear fragments in a dye density gradient, then used to infect monkey cells together with SV40 *tsA* as helper. Control cultures were infected with equivalent amounts of the recombinant or helper DNA alone. After 12 days at 41°, the cultures were freeze-thawed, and the resulting lysates were used to infect fresh monkey cell cultures. These were extracted 2 days later, and covalently closed circular DNA was purified and electrophoresed through 1% agarose gels. The following samples are shown from left to right: (1 and 5) covalently closed circular marker DNAs having lengths of (from the bottom of the gel) 2640, 3510, 4390, and 5000 bp. (2) Covalently closed circular DNA from cells infected with the recombinant alone lysate. (3) Covalently closed circular DNA from cells infected with the helper alone lysate. (4) Covalently closed circular DNA from cells infected with the recombinant plus helper lysate.

The 4570 bp recombinant species has been studied by DNA-DNA hybridization, DNA-RNA hybridization, electron microscopic heteroduplex analysis, and cleavage with 5 different restriction endonucleases. The results show that this species does contain the $tRNA^{Tyr}$ <u>su^{+}III</u> sequence, and that its structure has been faithfully preserved, at least to a resolution of about 50 bp.

Recently Fareed and Upcroft (this volume) have succeeded in propagating the SV40-su^{+}III recombinant in monkey cells without helper, by infecting with purified hybrid virus DNA then picking persistently infected clones. These cells appear to contain substantial quantities of intracellular recombinant DNA, but do not produce virus.

TRANSCRIPTION

The predominant forms of SV40 messenger RNA observed late in infection sediment at approximately 19 and 16s (reviewed in ref. 18). The 19s species, which is probably a precursor of the 16s species, extends from a 5' terminus close to the origin clockwise to a 3' terminus at about 0.2 map units. While the location of the promoter for this message is not known, it seems likely to lie outside of the late region. Consequently, one might expect that cells infected with a transducing virus such as the SV40-su^{+}III recombinant described above would transcribe the inserted foreign DNA due to read-through from the viral promoter.

Some preliminary evidence supporting this notion is shown in Fig. 3. Monkey cells were infected with the SV40-su^{+}III recombinant plus SV40 tsA helper lysate, incubated at 41° for 2 days, then pulse labeled with ^{3}H-uridine for 1 hour. Total cell RNA was extracted and sedimented through a sucrose gradient, and each fraction was tested for viral and bacterial-specific RNA by filter hybridization. As expected, the viral RNA sedimented at 19 to 16s. Two size classes of bacterial transcripts were observed. About half of the bacterial RNA sedimented as a sharp peak at 19s, the same position as the viral transcripts, whereas the remainder sedimented as a broad peak around 7s. In control experiments with RNA from mock and wild type SV40-infected cells, no hybridization to bacterial DNA was observed. Presumably the long bacterial transcripts result from initiation at a viral promoter, read-through of the bacterial DNA, and termination at the normal viral site; however, this has not been tested directly. The origin of the short transcripts is obscure. They might result from initiation at a viral site and termination at a bacterial site, initiation at a bacterial site and termination at a viral site, or initiation and termination within the bacterial DNA. Alternatively, they might represent degradation or processing products of the larger hybrid species.

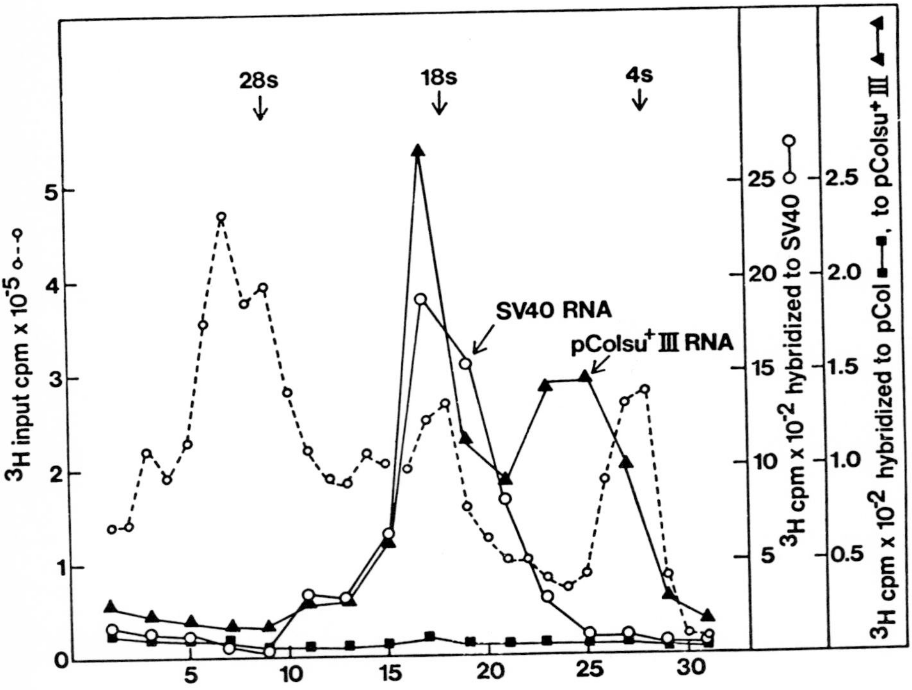

Fig. 3. Size distributions of viral and bacterial transcripts made in monkey cells infected with the SV40-*su*$^+$*III* recombinant. Monkey cells were infected with the SV40-*su*$^+$*III* recombinant plus SV40 *tsA* helper lysate, incubated for 2 days at 41°, then pulse labeled for 1 hour with ^{3}H-uridine. Total RNA was extracted, mixed with uniformly ^{32}P-labeled RNA as a marker, and sedimented through a sucrose gradient. An aliquot of each fraction was counted for ^{3}H and ^{32}P, and the remainder was hybridized with filters containing 1.0 μg of SV40 DNA, 2.9 μg of pCol DNA, and 2.9 μg of pCol*su*$^+$*III* DNA. (Plasmid pCol*su*$^+$*III* contains the *su*$^+$*III* fragment which was inserted into SV40, whereas the parental plasmid pCol does not.) The filters were incubated for 20 hours at 65°, then washed, treated with RNAase, washed, and counted. Under these conditions, the

efficiency of hybridization of various complementary RNAs to their template DNAs was 10%.

O---O Total ^{3}H-cpm x 10^{-5},
O——O ^{3}H-cpm x 10^{-2} hybridized to SV40 DNA,
Δ——Δ ^{3}H-cpm x 10^{-2} hybridized to pColsu$^+$III DNA,
—— ^{3}H-cpm x 10^{-2} hybridized to pCol DNA.

In the experiment shown in Fig. 3, the ratio of ^{3}H-labeled bacterial to viral transcripts was 10%, close to the ratio of 12% which would be expected if the two types of RNAs were synthesized and degraded at the same rates. More important, the ratio of ^{3}H-labled bacterial to total RNA was 0.6%. This ratio rose to 2% when the pulse time was shortened to 10 minutes. Thus, regardless of their origin, the bacterial transcripts represent a substantial portion of the total RNA synthesis in the productively infected monkey cells.

Attempts to detect mature suppressor tRNA in infected monkey cells were unsuccessful, suggesting that the mammalian cells are deficient in at least one of the several processes involved in tRNA biosynthesis in *E. coli*.

TRANSFORMATION

Since the SV40 late genes are not required for transformation (19), it seemed likely that it would be possible to stably transform non-permissive cells with recombinant viruses carrying foreign DNA in the late region. Would such transformants contain the foreign DNA?

To answer this, primary rat embryo cells were infected with purified SV40-su$^+$III recombinant virus DNA, and a pool of transformants was selected by subculturing in low serum medium. By the 5th passage, greater than 90% of the cells were positive for the SV40 t antigen. Total cell DNA was extracted from these pooled transformants and tested for viral and bacterial sequences by its ability to accelerate the reassociation of highly labeled SV40 restriction fragments and the *E. coli* su$^+$III fragment. These experiments showed that the transformants contain roughly 2 copies per diploid cell of the viral sequences included in the vector, but not of the viral sequences replaced by bacterial DNA. More crucially, they also contain about 2 copies of at least part of the su$^+$III fragment. Attempts to precisely map the bacterial sequences by the Southern method (20) are in progress (Fareed and Upcroft, this volume).

DISCUSSION

The SV40 late region deletion vehicle plus conditional early gene mutant helper system can be used to propagate foreign DNA fragments in both productively infected permissive monkey cells and in transformed non-permissive cells. The advantage of working in permissive cells is that the virus is present in many thousands of copies, thus facilitating biochemical studies. The advantage of working with non-permissive cells is that they are not killed by the virus, and can be grown indefinitely.

The cloning system is potentially quite general, since it is not necessary that the vector and the foreign DNA fragment have the same sort of termini. For example, a fragment having one *Eco*RI terminus and one *Bam*I terminus can be ligated to a vehicle with one *Eco*RI terminus and one *Hpa*II terminus. The progeny virus have one *Eco*RI site and a short deletion around the *Bam*I-*Hpa*II junction (D. Ganem, D.H. Hamer, and G.C. Fareed, unpublished results).

To date, SV40 has been used only to clone well characterized fragments of prokaryotic DNA. Clearly, the construction of viruses carrying eukaryotic genes would be of greater biological interest. Assuming that the inserted DNA is expressed, such viruses could be used to make functional maps of the *cis*-dominant regulatory sequences surrounding eukaryotic genes. This would provide a powerful complement to the methods which have been developed for cloning eukaryotic DNA in *E. coli*.

REFERENCES

(1) J.H. Miller, Experiments in Molecular Genetics (Cold Spring Harbor Laboratory, Cold Spring Harbor, New York, 1972).

(2) A.M. Campbell, Episomes (Harper and Row, New York, 1969).

(3) N.E. Murray and K. Murray, Nature, 251 (1974) 476.

(4) A. Rambach and P. Tiollais, Proc. Nat. Acad. Sci. USA, 71 (1974) 3927.

(5) M. Thomas, J.R. Cameron and R. Davis, Proc. Nat. Acad. Sci. USA, 71 (1974) 4579.

(6) D. Ganem, A.L. Nussbaum, D. Davoli and G.C. Fareed, Cell, 7 (1976b) 349.

(7) A.L. Nussbaum, D. Davoli, D. Ganem and G.E. Fareed, Proc. Nat. Acad. Sci. USA, 73 (1976) 1068.

(8) D.H. Hamer, Proceeding of the Tenth Annual Miles Symposium, in press.

(9) D.H. Hamer, D. Davoli, C.A. Thomas, Jr. and G.C. Fareed, J. Mol. Biol, in press.

(10) S.P. Goff and P. Berg, manuscript submitted to Cell.

(11) K.J. Danna and D. Nathans, Proc. Nat. Acad. Sci. USA, 69 (1972) 3097.

(12) G.C. Fareed, C.F. Garon and N.P. Salzman, J. Virol., 10 (1972) 484.

(13) P. Tegtmeyer, J. Virol., 10 (1972) 591.

(14) J.Y. Chou and R.G. Martin, J. Virol., 13 (1974) 1101.

(15) C.J. Lai and D. Nathans, Cold Spring Harbor Symp. Quant. Biol., 39 (1974) 53.

(16) T.E. Shenk, C. Rhodes, P.W.J. Rigby and P. Berg, Cold Spring Harbor Symp. Quant. Biol., 39 (1974) 61.

(17) B. Hirt, J. Mol. Biol., 26 (1967) 365.

(18) N. Acheson, Cell, 8 (1976) 1.

(19) F.L. Graham, P.J. Abrahams, C. Mulder, H.L. Heijneker, S.O. Warnaar, F.A.J. de Vries, W. Fiers and A.J. Van der Eb, Cold Spring Harbor Symp. Quant. Biol., 39 (1974) 637.

(20) E.M. Southern, J. Mol. Biol., 98 (1975) 503.

The work described in this paper was performed in collaboration with D. Davoli, C.A. Thomas, Jr., G.C. Fareed, and G. Khoury, and was supported by research grants from the National Institutes of Health.

Discussion

U. Littauer, The Children's Hospital of Philadelphia: Have you attempted to process in vitro the SV40-SU$^+$ III transcript with E. coli enzymes and generate mature tRNA, similar to the experiments of V. Daniel, who transcribed in vitro ϕ 80 SU$^+$ III DNA and then processed the product to obtain biological active tRNA?

D. Hamer, Harvard Medical School: No, we didn't try to do that though it is probably a good idea. To get enough tRNA to put into a protein synthetase assay would require too many monkey cells. Hopefully, Ravi Darr, who is working in George Khoury's lab, is going to try to purify those transcripts by hybridization and then look directly at their sequence.

P. Berg, Stanford University Medical Center: Dean, I wonder if you have been unduly pessimistic about the results of your assays for the tRNA? What do you think is the smallest amount of tyrosine tRNA you could have detected in the infected monkey cell RNA preparation. For example, did you perform reconstructions with varying amounts of E. coli tyrosine tRNA added to the infected monkey cell RNA to see how small an amount of that acceptor activity you could detect.

D. Hamer: That is exactly what I did. Our experiment showed that suppressor tRNA could be detected at a level of 1 to 2 parts in a thousand.

P. Duesberg, University of California: I have what is perhaps a trivial question. You said that the ratio of E. coli RNA and SV40 RNA made in the infected cell was expected to be 10 to 1. Does that take into consideration that you have helper virus present?

D. Hamer: Yes, a factor in that calculation is that you have 70% helper virus and only 30% recombinant virus.

P. Deusberg: Doesn't the helper virus outgrow the hybrid recombinant DNA?

D. Hamer: No, the recombinant is not overgrown. The ratio of recombinant to helper remains the same after an additional passage.

P. Duesberg: Could I ask you a last question? You said the recombinant DNA is made in the rat cell but you do not get virus production. Could you make a hybrid between SV40 and polyoma and determine which DNA segments control virus production in permissive and non-permissive cells? You could make an SV40 polyoma hybrid and plate or mouse or monkey cells; or is that verboten by the NIH guidelines?

D. Hamer: I am not sure. Probably Dr. Berg would know. I guess that is a P-4 experiment. It certainly is feasible because the restriction maps and the markers are known well enough.

P. Duesberg: That has not been done?

D. Hamer: Not that I know of.

recombinant DNA preparation showed that the region of homology between the λ immunity region fragment and the recombinant DNA was preserved in this hybrid (Fig. 2).

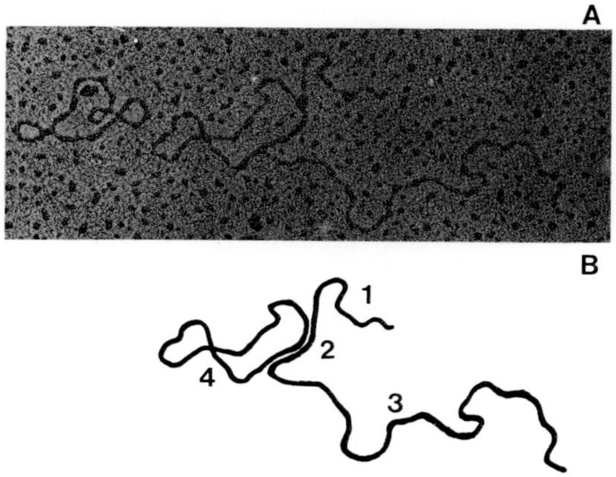

Fig. 2. Electron Microscopic Heteroduplex Analysis of Chimeric (λ-d)$_3$ Genomes.

The heteroduplex between open circular (λ-d)$_3$ and linear λ-EcoRI-B was prepared and measurements of 25 molecules were made. The region of homology between (λ-d)$_3$ and λ-EcoRI-B (designated 2 in the sketch) is 8.3% of the length of the λ-EcoRI-B segment and begins 21.7% from one end of λ-EcoRI-B (designated 1) and 70.0% from the other end (designated 3). The position of the duplex region in the heteroduplex is as predicted from the location of the λ 520 base pair segment within λ-EcoRI-B (Figure 3). The sum of segments 2 and 4 is approximately 83% the length of wild-type SV40 DNA (1.54 μm). (Reprinted, with permission, from Ganem *et al.*, 1976, Cell *7*, 349)

This was substantiated also by hybridization analysis where the 1400 base pair fragment and the trimeric (λ-d)$_3$ hybrid both annealed specifically with denatured λ DNA immobilized on nitrocellulose filters.

To establish the generality of this method and the feasibility for molecular cloning of hybrid genomes of this type we focussed our attention on a larger neighboring segment from the λ immunity region which included the leftward operator promoter site, as shown in Fig. 3. This was excised as a 2400 base pair segment bounded by BamI and HindIII

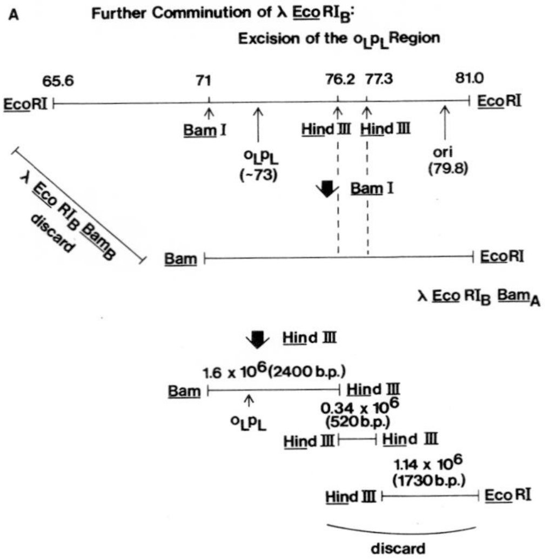

Fig. 3. Preparation of DNA fragments used for in vitro recombination. Schematic representation of the excision of the λ DNA fragment carrying the leftward operator, O_LP_L, from the immunity region located in λ-R.EcoRI fragment B.

termini and purified by gel electrophoresis. The molecular cloning of this hybrid (7) in mammalian cells not only showed the efficacy of defective SV40 vectors, but also provided an opportunity for our examining a well-characterized prokaryotic genetic regulatory site, the leftward operator, in a eukaryotic environment. This region of λ seemed appropriate for the construction of novel recombinant DNA molecules, since it lacked the replication functions of λ and the leftward operator that it contained could be specifically identified in a hybrid molecule by λ repressor binding. The vector used to clone this λ segment was excised from a triplication mutant by EcoRI and BamI cleavages and was a fragment of the monomer segment from this reiteration mutant. EcoRI cleavage of this DNA generated the one-third wild-type SV40 monomer, whose physical structure had been previously determined (8). Subsequent BamI cleavage produced a 940 base pair segment bearing the origin and a smaller 730 base pair piece. We linked this vector segment via the BamI end to the corresponding terminus of the λ 240 base pair segment. After mixed infection of monkey cells with this ligation mixture and wild-type DNA, the progeny viral DNA was found to contain specific λ sequences by filter hybridization. This hybrid DNA was enriched with an infectious center plaquing procedure in which cells infected with helper and defective hybrid genomes are seeded onto a monolayer of uninfected cells. The single infectious center plaque out of about 40 tested for λ DNA sequences was subsequently used to generate a virus stock. The progeny intracellular DNA from this stock contained two different shortened genomes in addition to wild-type SV40 DNA as judged by agarose gel electrophoresis (Fig. 4).

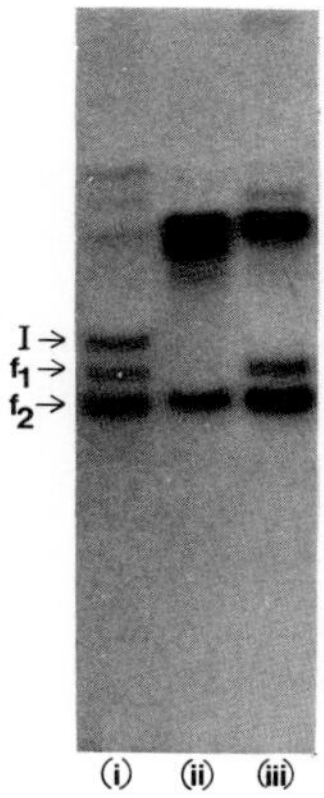

Fig. 4. Properties of the DNA derived from plaque i.c. 219. The infectious center method using DNA infection rather than virus infection was employed to clone the λ-SV40 hybrid. A DNA infection was carried out as follows: DNA (primarily viral) was extracted from cells infected with the original lysate created after infection of monkey cells with the recombinant DNA and WT SV40 DNA, and shortened viral DNA (1.2 μg) was purified by 1.4% agarose gel electrophoresis. This DNA, mixed with 0.6 μg of WT SV40 DNA, was used to infect a confluent monolayer of CV-1 cells in a 100 mm culture dish. After 18 hr, the cells were trypsinized and counted. Appropriate volumes of the diluted cell suspensions were seeded into 60 mm culture dishes along with about 2 x 10^6 uninfected cells. After 24 hr at 37°, each dish was covered with a 1% agar overlay in Eagle's medium with 2% fetal calf serum, and additional overlays were made after 5, 9, and 12 days at 37° (the last one containing 0.01% neutral red). Plaques were readily evident at dilutions of 10^{-2} and 10^{-3} of the original infected cell suspension. These plaques were aspirated (0.2 ml) and 0.1 ml of each was used to infect a fresh culture of CV-1 cells in a 35 mm dish. After labeling with [^{3}H]thymidine, viral DNA was selectively

Fig. 4 (cont.) extracted and tested for hybridization to λ DNA on filters. One plaque aspirate, i.c. 219, proved to be positive and the remainder of the aspirate from this plaque was used to infect monkey cells in a 150 mm dish. The lysate resulting from this infection provided the viral stock for preparing intracellular viral DNA analyzed above. This shows an autoradiogram of i.c. 219 Form I (closed circular, superhelical) [^{32}P]DNA which had been subjected to electrophoresis in a 1.4% agarose slab gel before (i) and after treatment with R.BamI (ii) or R.EcoRI (iii).

When these superhelical genome species were separately tested for homology to the λ immunity region fragment, the shortest genomes, designated f2, annealed to the greatest extent. To be more certain that the hybrid genomes retained the leftward operator sequences, we next examined the affinity of λ repressor for these DNAs. Not surprisingly, as shown in Fig. 5, the

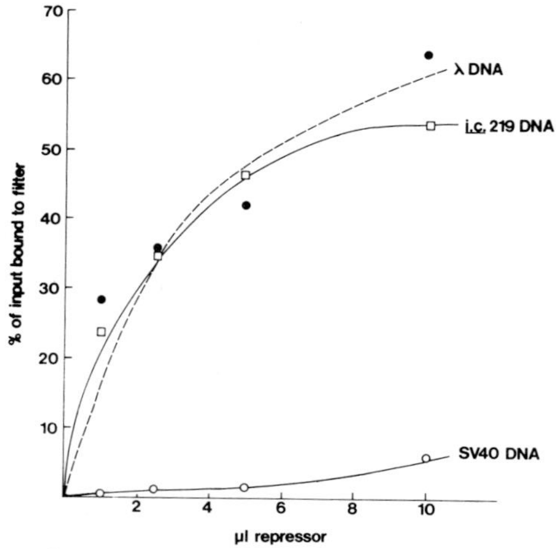

Fig. 5. Affinity of λ repressor for i.c. 219 DNA. Viral DNA was exposed to increasing amounts of λ repressor protein and passed through nitrocellulose filters. Material retained by filters is expressed as percent of input radioactivity. The binding assay mixture (0.6 ml) contained 0.02 μg of i.c. 219-f2 [^{32}P]DNA (10,000 cpm) and either λ [^{3}H]DNA (0.25 μg; 5000 cpm) or SV40 [^{3}H]DNA (0.3 μg; 5000 cpm) and the indicated volumes of λ repressor (3 x 10^{-8} M). Binding of f2 by λ repressor was abolished by addition of nonradioactive λ DNA in excess.

f2 species bound λ repressor at levels where negligible binding was detected to wild-type SV40 or the f1 species. This observation was further refined by our asking whether λ repressor could selectively bind a specific restriction endonuclease cleavage fragment from the chimeric DNA. _HpaII_ cleavage of the λ immunity region is known to produce two small fragments of 350 and 570 base pairs in addition to other fragments. Maniatis and Ptashne have shown that the leftward operator is located on the 350 base pair fragment (15). When the chimeric DNA was cleaved by _HpaII_ and allowed to react with λ repressor, the 350 base pair fragment was bound by repressor. The gel electropherogram of the DNA fragments bound by λ repressor and subsequently eluted from a nitrocellulose filter by SDS treatment is shown in Fig. 6.

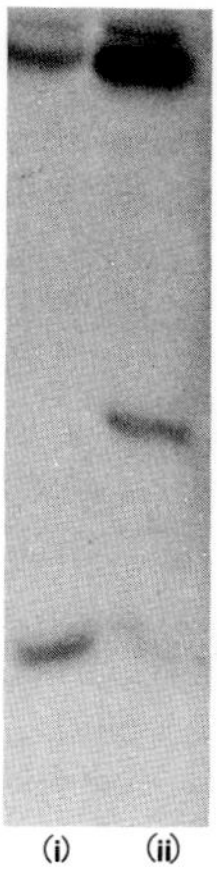

Fig. 6. λ Repressor binds the 350 bp fragment from the R._HpaII_ digest of _i.c._219-f2 [^{32}P]DNA. Unfractionated _i.c._219 DNA (2.3 x 10^6 cpm, 12.4 μg) was cleaved with R._HpaII_ and concentrated by alcohol precipitation in the presence of 200 μg of calf thymus DNA. It was then exposed to repressor protein (10^{-10} M) in a total volume of 2 ml. The resulting complex was retained on a filter and eluted with a buffer containing sodium dodecyl sulfate. A 10% aliquot of the eluate (7000 cpm) was subjected to electrophoresis and autoradiography (i). (ii): _i.c._219-f2 DNA was cleaved with R._HpaII_. The 3000 bp segment in i was retained, probably due to nonspecific binding to the filter, whereas of the two small fragments only the 350 bp fragment was retained.

In channel 2 is the _HpaII_ cleavage of f2 DNA which reveals the two small fragments and a large 3000 base pair fragment which contains the SV40 vector segment. In 1 is shown the DNA retained by λ repressor. Of the two small fragments, only the

350 base pair fragment was bound as predicted. These structural findings were confirmed by electronmicroscopic heteroduplex analysis. The hybrid DNA was nicked, denatured, and allowed to anneal with denatured λ EcoRI immunity region fragment. The homology between the two DNAs was seen to be about 32% of the λ fragment, or 2300 base pairs, and was appropriately situated in the λ fragment to cover the region of the leftward operator. To map more precisely the λ segment in the hybrid, the λ EcoRI-B immunity region DNA fragment was cleaved by endo R.HindIII, as seen in Fig. 7.

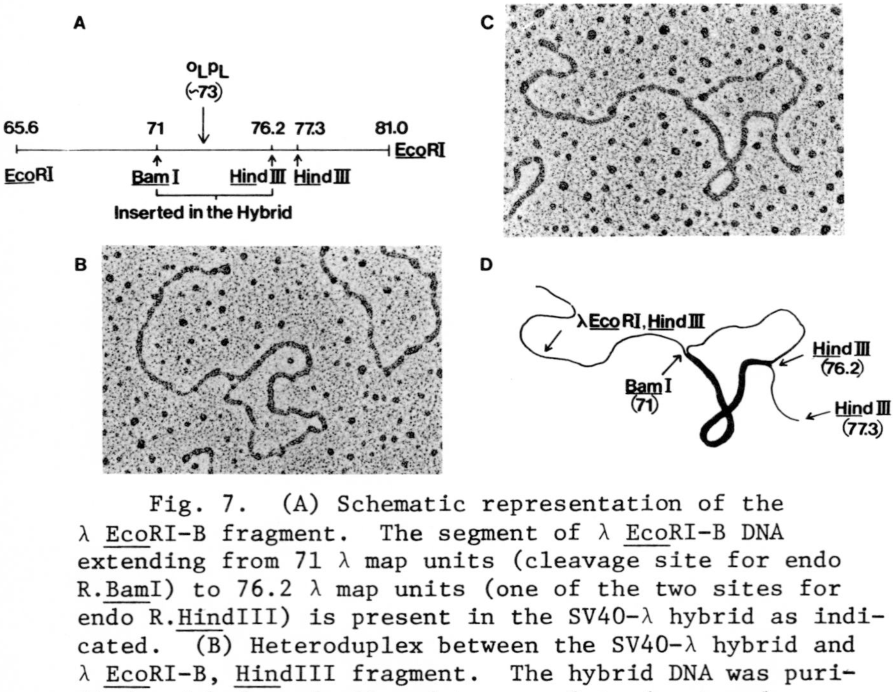

Fig. 7. (A) Schematic representation of the λ EcoRI-B fragment. The segment of λ EcoRI-B DNA extending from 71 λ map units (cleavage site for endo R.BamI) to 76.2 λ map units (one of the two sites for endo R.HindIII) is present in the SV40-λ hybrid as indicated. (B) Heteroduplex between the SV40-λ hybrid and λ EcoRI-B, HindIII fragment. The hybrid DNA was purified, nicked, and allowed to anneal to denatured λ EcoRI-B (further cleaved with endo R.HindIII). A partially duplex region is located in the circular portion of this representative heteroduplex molecule. The presence of one linear single-stranded segment extending from the circle proves that the hybrid DNA contains that portion of λ EcoRI-B up to the HindIII cleavage site at 76.2 map units. The single-stranded linear tail measures 0.359 ± 0.02 λ EcoRI-B lengths as expected for the part of the EcoRI-B, HindIII fragment between 65.6 and 71 λ map units that is not present in the hybrid genome. (C) Heteroduplex between the hybrid and λ EcoRI-B,HindIII (incomplete cleavage). A few heteroduplexes of this type

Fig. 7 (cont.) were seen which probably were created from annealing of a hybrid molecule with the λ *EcoRI*-B fragment cleaved only once by *Hind*III at 77.3 map units. (D) Diagrammatic representation of the heteroduplex shown in (C), with the duplex region of homology drawn as a thick line. (Reprinted, with permission, from Davoli *et al.*, 1976, J. Virol. *19*, 1100)

This produces three fragments, one of which, that spanning 65.6-76.2 λ map units, should form a heteroduplex with the hybrid. Such heteroduplexes were observed with one single-stranded branch extending from the circular region. A few heteroduplexes formed between the hybrid and the λ fragment, which had been cleaved only once by *Hind*III at 77.3 λ map units. These results indicated that no gross rearrangements of a region of λ DNA between 71 and 76 map units had occurred in the SV40-λ hybrid.

The final structural question concerned the SV40 vector sequences in the hybrid. Since the hybrid genomes were present in at least two-fold molar excess to helper genomes in the original virus stock, we anticipated that at least part of the vector sequences, including the replication origin, had been duplicated. This was substantiated (9) by an analysis of restriction endonuclease fragments known to arise from the vector segment and by the following heteroduplex analysis, as illustrated in Fig. 8. In the heteroduplex analysis the *Hpa*II 3000 base pair linear fragment from the hybrid genome was denatured and annealed to single-stranded circular molecules of the triplication mutant from which the vector had been obtained. The heteroduplexes, such as that shown here, contained two regions of homology separated by a small deletion loop. These regions of homology indicate a tandem duplication of most of the vector sequences in the hybrid. This duplication occurred *in vivo* during the propagation of this hybrid. It provided two advantages: (1) the size of the hybrid genomes was increased to allow for efficient encapsidation since viral molecules less than 70% of wild-type size are not packaged, and (2) a selective advantage in DNA replication due to the duplicated origin for SV40 DNA replication.

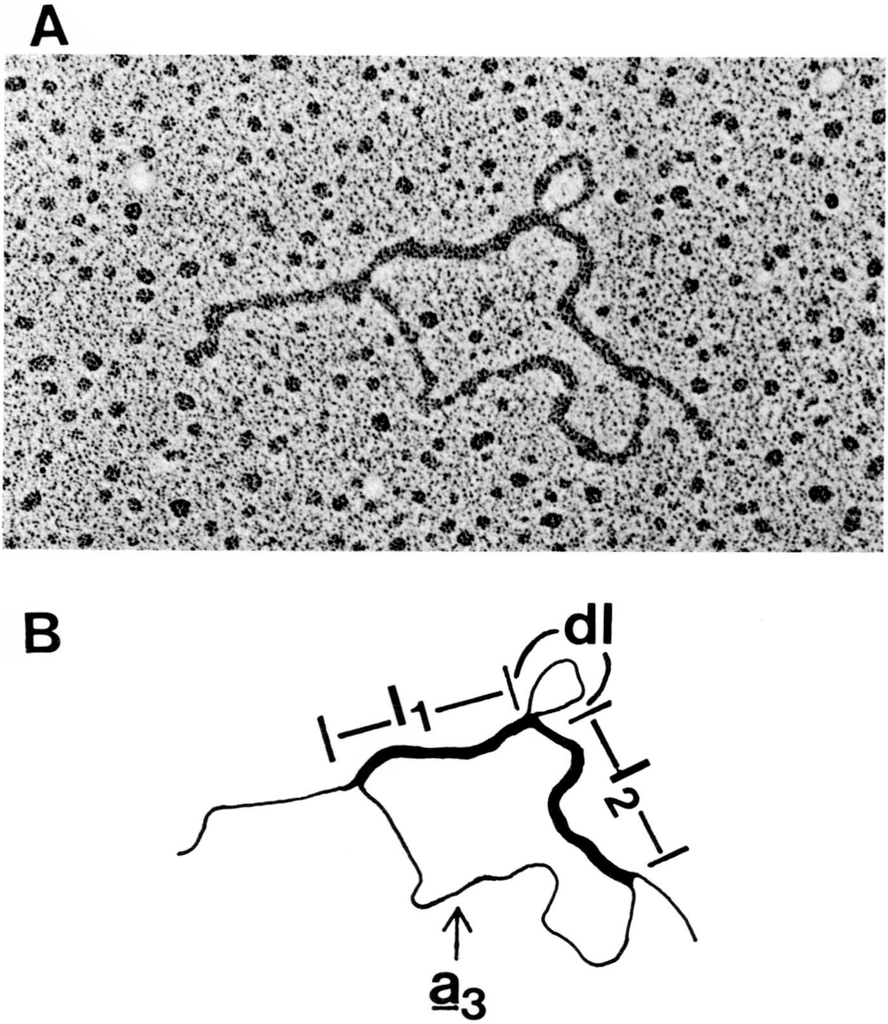

Fig. 8. Heteroduplex between reiteration mutant $\underline{a}_3$ and the HpaII-3000-base pair fragment. Reiteration mutant $\underline{a}_3$ DNA I was nicked and then denatured and renatured in the presence of the HpaII-3000-base pair fragment from the SV40-λ hybrid. This fragment had been purified after endo R.HpaII cleavage, as described in the legend to Fig. 6, and electron microscopy analysis was carried out. Measurements were made of 15 heteroduplexes, using single-stranded and duplex molecules of $\underline{a}_3$ as standards. The two duplex regions (shown as a heavy line in the diagram [B] and lettered I_1 and I_2), which

Fig. 8 (cont.) are 0.184 ± 0.006 and 0.182 ± 0.006 of the a_3 genome, respectively, represent a tandem duplication of the majority of the vector segment aBam-A preserved in the hybrid. The small, single-stranded deletion loop (dl in part B) is 0.160 ± 0.007 of the a_3 genome and is that part of fragment a not present in the hybrid. The single-stranded tails (thin lines in part B) are the remaining segments of λ DNA included in the HpaII-3000-base pair fragment. (Reprinted, with permission, from Davoli *et al.*, 1976, J. Virol. 19, 1100)

These findings with small segments from phage λ DNA and short vector segments from SV40 reiteration mutants indicated that defective SV40 replicons could serve as vectors for molecular cloning and propagating foreign DNA in mammalian cells. However, since hybrid genomes constructed with non-complementing SV40 replicons, such as those used in these experiments, must be propagated and cloned with wild-type helper genomes, there is no simple means for selection for the hybrid genomes in a mixed population. To overcome this deficit, vector segments containing intact early or late gene regions could be employed. For example, excision of a suitable fragment from the late gene region of SV40 and insertion of a prokaryotic segment of similar size would create a defective hybrid capable of complementing early (tsA) mutants. In collaboration with Dean Hamer and Charlie Thomas at Harvard Medical School, we recently employed this approach to clone a bacterial transfer RNA suppressor gene in monkey cells (10). The basic features of the construction, propagation and biological characterization of one such complementing recombinant genome are reviewed by Hamer *et al*. (this symposium). The results of our findings and those of Goff and Berg (11) now have shown that deletions in the B, C and D regions of SV40 of 1300 base pairs from the HpaII site at .74 map units to the EcoRI site at 0 map units and of 2000 base pairs from the HpaII site to the BamI site at .15 map units can be replaced by foreign DNAs to produce genomes of over 90% of the original length. We had shown in studies of reiteration mutants that genomes of lengths of less than about 70% of SV40 size or greater than 100% of wild-type size were not encapsidated in progeny virions. The bacterial nonsense suppressor gene we inserted into SV40 is su^+III from *E. coli*. This gene specifies a tRNA which translates the amber termination codon UAG as tyrosine. The source of the bacterial DNA was an *E. coli* plasmid constructed by Hamer called pCol-su^+III which carries a single copy of the tRNA tyrosine su^+III gene. Cleavage of this plasmid with the site-specific restriction endonucleases EcoRI and HpaII generated an 860 base pair fragment which included the suppressor tRNA structural gene sequence and its promoter and transcription termination regions. The viral vehicle was a 3700 base pair fragment of wild-type SV40 DNA and also produced by cleavage with EcoRI and HpaII. Since both EcoRI and HpaII produced DNA molecules with unique cohesive termini, the su^+III and SV40 fragments could anneal with one another in a single orientation and could then be covalently joined by treatment with polynucleotide ligase. This procedure yielded circular recombinant molecules which are 92% the length of wild-type SV40 DNA, and the su^+III fragment was inserted in such a way that the 5' end of the tRNA structural gene was proximal to the 5' end of the 19S SV40 late region transcript.

The SV40-su^+III recombinant DNA molecules were purified by two cycles of dye-density gradient centrifugation and, together with helper DNA from SV40 with a temperature sensitive mutation in gene A, were used to infect monkey cells at 41°C, which is nonpermissive for the helper. Under these conditions, virus was produced only by cells infected with both recombinant, which provides the early function, and the helper, which provides the late functions. Both the recombinant and helper genomes were encapsidated into virions, and this generated a mixed lysate which could subsequently be used to infect monkey cells for the preparation of virus-specific DNA and RNA. An analysis of the covalently closed circular viral DNA preparations from cells infected with the recombinant plus helper lysate by agarose gel electrophoresis showed two prominent viral species; about 70% of the viral DNA migrated at the position of helper, 5000 base pairs, and the remainder migrated at exactly the position expected for the SV40-su^+III recombinant, 4580 base pairs.

Transfection of Rat and Monkey Cells by the SV40-*E. coli* Hybrid DNA

Because of the conservation of the origin of replication and the A gene function, the hybrid SV40-*E. coli* DNA (SV40-su^+III) should be capable of transformation of non-permissive cells and of autonomous replication in permissive cells. We have transformed secondary rat embryo cells with the purified SV40-su^+III DNA and constructed long term persistent infections of TC7 monkey kidney cells with this chimera. Various restriction enzymes, the Southern (12) elution procedure to transfer separated DNA segments from agarose gels to nitrocellulose filters and hybridization with probes specific for SV40 and $tRNA^{Tyr}su^+III$ have been used to analyze the fate of the recombinant DNA in both host cell systems.

Hybrid DNA in Transformed Rat Embryo Cells

Digestion of total cellular DNA from cloned rat embryo cells with BalI (an enzyme which does not cleave SV40) and hybridization with the nick-translated SV40 probe (Fig. 9) reveal two bands migrating in the vicinity of linear SV40 (EcoRI-generated), one of which migrates very close to SV40 DNA III. The other species is larger, and could possibly correspond to open circles of the linear. However, digestion of total cellular DNA with EcoRI or BamI, which each cleave the original SV40-su^{+}III hybrid once, leave both these bands in the same proportion as the Bal digest. Furthermore, cleavage with SalI, another enzyme which does not cleave the hybrid, or with BalI+Sal reveals both DNA species (data not illustrated). Two more high molecular weight DNA bands are seen near the top of the gel for the BalI digest. These probably correspond to integrated SV40-su^{+}III since cleavage with EcoRI or BamI does not show resistant oligomers in the corresponding region of the gels, although a number of species is seen which migrate between the two pairs of bands. Cleavage with a vast excess of SalI, sufficient to cause conversion of supercoiled SV40, by nicking, to >50% open circles, under similar migration conditions, still reveals the same two bands, and only these two bands. The two lower bands hybridize the nick-translated su^{+}III tRNA gene probe, and the simplest interpretation is that they correspond to one (or, possibly, two) free viral DNA species.

Cleavage with EcoRI or BamI and hybridization with the SV40 probe produces a considerable number of bands, including the two revealed in DNA cleaved by BalI and SalI. One extra prominent band is common to both digests. Another prominent band is seen in the BamI-cleaved sample. Both bands migrate faster than the SV40 marker, and do not appear to be digestion products of the two most prominent bands because the proportion and amount of the latter are similar in each digest. Furthermore, they do not appear to be another masked free viral DNA species because they were not generated during the excess SalI digestion. If the two bands common to all cleavage patterns were integrated DNA, one would expect their migration to change under the different enzyme digestions. Their conservation argues for their being free viral DNA. We do not know, however, if the free viral DNA is common to a few producer cells or is distributed equally among all cells.

A number of other species are generated by EcoRI and BamI cleavages, some migrating slower than SV40, most of which are common to both digests and probably are enzyme resistant oligomers of the free viral input DNA. Those which migrate

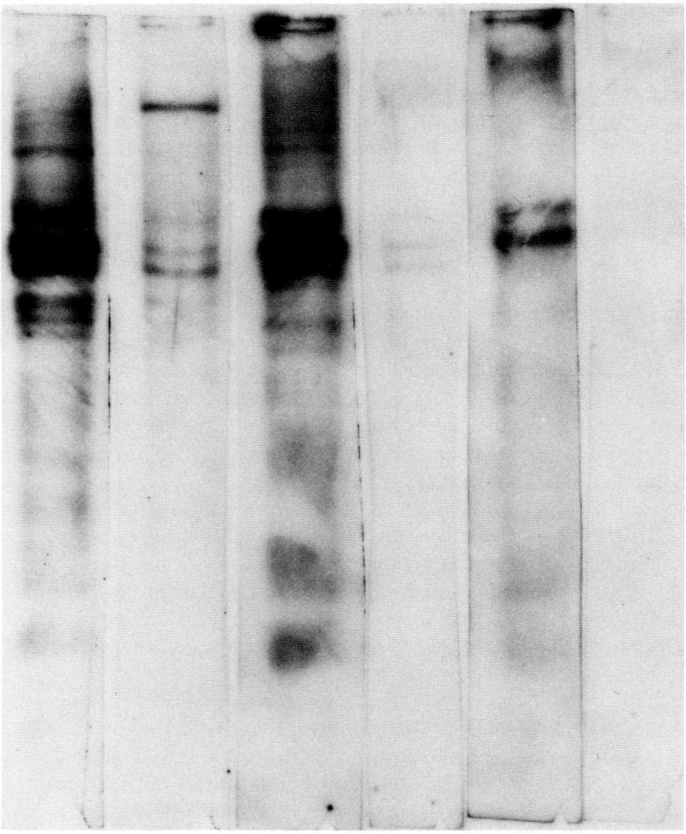

Fig. 9. Hybridization analysis of cloned secondary rat embryo cells transformed by SV40-su$^+$III. A 20 μg sample of total cellular DNA, prepared by the method of Ketner and Kelly (13) was subjected to electrophoresis in 1% agarose in a horizontal slab gel apparatus (Upcroft et al., manuscript in preparation). Each channel was excised and eluted onto nitrocellulose filters as described by Southern (12). Each filter was hybridized with either nick translated (16) SV40 DNA or pColsu$^+$III DNA at a specific activity of 5 x 10^7 cpm/μg and autoradiography was performed (Upcroft et al.). Gel slots (from left to right) were loaded with the following: (nos. 1 and 2) EcoRI-cleaved DNA, (nos. 3 and 4) BamI-cleaved DNA, (nos. 5 and 6) BalI-cleaved DNA. Nos. 1, 3 and 5 were hybridized with the SV40 probe and 2, 4 and 6 with the pColsu$^+$III probe. Exposure time was 60 h.

faster than SV40 are not identical in the two restriction nuclease products. These latter bands hybridize the tRNA gene probe and we interpret these to be the products of the integrated SV40-su$^+$III genome (Fig. 10). They are not generated by simple cleavage of the two major bands common to all the cleavage patterns illustrated and their yield is too great to be products of higher oligomers.

One very conspicuous high molecular weight species ($>10^7$) appears after hybridization of the EcoRI digest with the tRNA probe. It migrates in the vicinity of high molecular weight species identified with the SV40 probe, but its intensity is too great to be a simple oligomer of the SV40-su$^+$III hybrid. Furthermore, the band does not appear in BamI cleavage patterns. We interpret these data to mean that this band may be a result of integration of the hybrid DNA near the HpaII site of one end of the suppressor insert. Cleavage of the cellular DNA by EcoRI would be likely then to separate the bacterial and SV40 sequences onto different DNA fragments. Hybridization would then yield a species which would hybridize poorly to the SV40 probe, but strongly to the tRNA gene probe. Note that one additional band of lower molecular weight which hybridizes to both probes, is created by BamI cleavage. The BamI site is removed from the EcoRI site on the SV40 genome by 14% and leaves considerably more SV40 sequences which may hybridize.

Hybrid DNA in Persistently Infected TC7 Cells

Analysis of total cellular DNA from a cloned line of TC7 cells persistently infected with the SV40-su$^+$III DNA by ethidium bromide staining of agarose gels showed three major superhelical DNAs, and a minor fourth, the second smallest of which was predominant and comigrated with SV40 DNA. Similarly, four open circles were present in lower amounts. By the Southern procedure, each of these hybridized to a nick-translated SV40 probe, the second species hybridizing the most SV40 nick-translated DNA probe (Fig. 11). Higher molecular weight species, probably dimers, were also seen. Cleavage by EcoRI, followed by hybridization with the SV40 probe showed an increase in proportion of the second and third species but with all three superhelical molecules still remaining. Cleavage with BamI removed all detectable superhelical DNA and considerably increased the production of the smallest linear, as well as the second and third. A new species was generated, in low yield, migrating between supercoils and linears. The fragments of higher molecular weight produced by either BamI or EcoRI (detected as faint bands above the major bands) probably arise from integrated copies of the hybrid DNA. All four linear or open circular species plus one of higher molecular

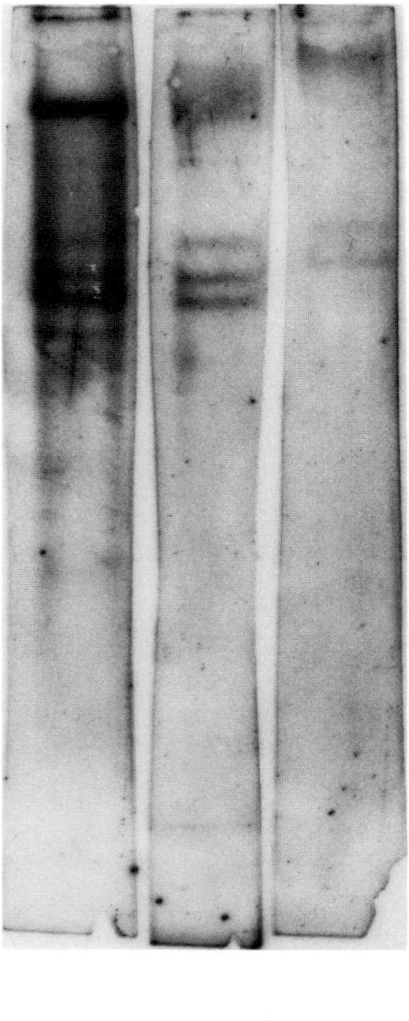

Fig. 10. Hybridization analysis of transformed rat embryo cells. This autoradiograph is a one week exposure of gel slots 2, 4 and 6 of Fig. 9 (i.e., EcoRI-, BamI- or BalI-cleaved DNA, hybridized with the pColsu$^+$III probe).

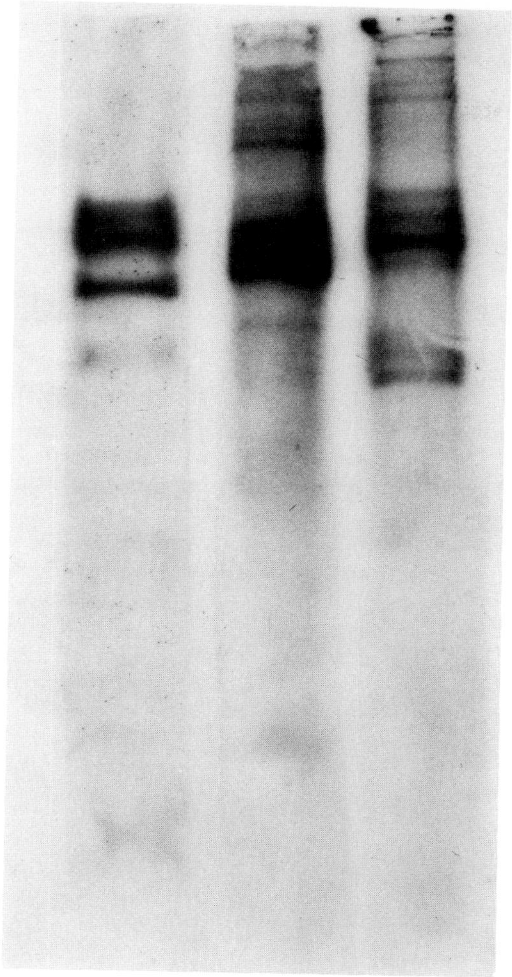

Fig. 11. Hybridization analysis of hybrid viral DNA from a persistently infected monkey cell line. The purified SV40-su$^+$III DNA was used to transfect TC7 monkey kidney cells and a chronically infected cell population was subsequently cloned (Fareed and Upcroft, in preparation). The total cellular DNA from this culture was purified and analyzed in the Southern (12) technique after cleavage by endo R.EcoRI (autoradiograph on right) or endo R.BamI (center autoradiograph). The hybridization probe was ^{32}P-labeled cRNA from SV40 DNA (13). The autoradiograph on the left shows the EcoRI-cleaved input SV40-su$^+$III DNA (note that rearrangements had occurred in this preparation during virus propagation).

weight were found to hybridize with the tRNA probe.

CONCLUDING REMARKS

These experiments thus show that SV40 vectors are efficient for the propagation of specific prokaryotic DNA segments in mammalian cells. The clear advantage of a complementing vector, such as one which has a deletion in the late gene region, is that the hybrid recombinant can be easily cloned with a complementation technique. Furthermore, once the hybrid DNA has been purified free of the helper DNA, one can transform a variety of mammalian cells with it. As found in this work, both vector and bacterial sequences remain associated with transformed rat embryo cells and persistently infected monkey kidney cells. An unexpected observation with the rat cells has been the presence of both free or unintegrated and integrated hybrid viral DNA sequences. Both of these host cell-vector systems have two advantages over the lytic infectious system for SV40: (1) no infectious virus is produced and (2) the size of the non-viral (prokaryotic or eukaryotic) DNA inserted in the late gene region of the vector may be quite large (no requirement for encapsidation in progeny virus).

ACKNOWLEDGEMENTS

This work was supported in part by a research grant from the National Cancer Institute, USPHS No. CA 20794-01. The expert technical assistance of Hagit Skolnik is gratefully acknowledged.

REFERENCES

(1) K. Struhl, J. Cameron and R.W. Davis, Proc. Nat. Acad. Sci. USA, 73 (1976) 1471.

(2) D. Ganem, A.L. Nussbaum, D. Davoli and G.C. Fareed, J. Mol. Biol., 101 (1976) 57.

(3) D. Davoli and G.C. Fareed, Nature, 251 (1974) 153.

(4) T. Shenk and P. Berg, Proc. Nat. Acad. Sci. USA, 73 (1976) 1513.

(5) D. Ganem, A.L. Nussbaum, D. Davoli and G.C. Fareed, Cell, 7 (1976) 349.

(6) D. Davoli, D. Ganem, A.L. Nussbaum, G.C. Fareed, P. Howley, G. Khoury and M.A. Martin, Virology, in press March 1977.

(7) A.L. Nussbaum, D. Davoli, D. Ganem and G.C. Fareed, Proc. Nat. Acad. Sci. USA, 73 (1976) 1068.

(8) G. Khoury, G.C. Fareed, K. Berry, M.A. Martin, T.N.H. Lee and D. Nathans, J. Mol. Biol., 87 (1974) 289.

(9) D. Davoli, A.L. Nussbaum and G.C. Fareed, J. Virol., 19 (1976) 1100.

(10) D. Hamer, D. Davoli, C.A. Thomas, Jr. and G.C. Fareed, J. Mol. Biol., in press 1977.

(11) S. Goff and P. Berg, Cell, in press 1977.

(12) E. M. Southern, J. Mol. Biol., 98 (1975) 503.

(13) G. Ketner and T.J. Kelly, Jr., Proc. Nat. Acad. Sci. USA, 73 (1976) 1102.

(14) M. Botchan, W. Topp and J. Sambrook, Cell, 9 (1976) 269.

(15) T. Maniatis and M. Ptashne, Nature, 246 (1973) 133.

(16) T. Maniatis, S.G. Kee, A. Efstratiadis and F.C. Kafatos, Cell, 8 (1976) 163.

Discussion

P. Berg, Stanford University: How many of the SV40 genome are there in the TC7 lines and in the rat embryo?

G. Fareed, Harvard Medical School: In the TC7 line we have not determined that by reassociation kinetic analyses. It is a large number of copies but we just have not quantitated that and Hamer will describe the reassociation kinetic studies of the transformed cells this afternoon. If you wait until then you will learn that information.

D. Nathans, Johns Hopkins University School of Medicine: Do you get polymers of your SV40 hybrid DNA?

G. Fareed: Yes. There is a large proportion of monomers and a small proportion of multimers; it is primarily monomer molecules that we have identified after many generations of these persistently infected cells.

D. Nathans: It was not clear to me that you have integration in those cells. I was wondering what stays up near the top of the gel. Could it be multimers in those transfer experiments that you did?

G. Fareed: It could be oligomeric hybrid viral DNA. Are you asking what the hybridization analysis detected in the slight amounts of material near the top of the gel?

D. Nathans: Yes.

G. Fareed: I think it suggests that there is a low proportion of multimers in the total cellular DNA.

D. Nathans: I was wondering about the evidence for integration in those persistently infected cells.

G. Fareed: The evidence for integration into cellular DNA is indirect at present. This evidence comes from the EcoRI or BamI cleavage of that total cellular DNA which generates, in addition to the form II and form II species of the free hybrid DNA, bands of high molecular weight. These could arise from integrated forms of the viral DNA.

J. Sambrook, Cold Spring Harbor Laboratory: Jim McDougan, Michael Botchan and myself have been looking at a set of clones of SV40-transformed human cells with properties similar to those described by Dr. Fareed. They contain large amounts of "free" viral DNA, in numbers ranging up to 1000 molecules per cell. When such cultures are examined by in situ hybridization it becomes clear that most of the viral genomes are produced by a small number of the cells. So there is a "jackpot" effect. We do not yet know whether or not the cells contain integrated viral DNA sequences.

G. Fareed: Very interesting. We plan to use in situ hybridization with the SV40 and suppressor plasmid probes to examine this possible "jackpot effect" in the rat cells transformed with the SV40-SU^{+}III hybrid.

HR-T MUTANTS OF POLYOMA VIRUS

THOMAS L. BENJAMIN
Department of Pathology
Harvard Medical School
25 Shattuck Street
Boston, Massachusetts 02115

INTRODUCTION

Neoplastic cell transformation by polyoma virus or SV-40 is accompanied by the integration of the viral genome of $3x10^6$ daltons into the host cell genome of about $3x10^{12}$ daltons. A variety of evidence - genetic, radiobiological, and nucleic acid hybridization - makes it clear that while the entire viral genome may persist, the expression of only a portion of it is vital for transformation to occur. This portion corresponds to the half of the genome transcribed early in a lytic infection. Transformation is also accompanied by changes in cellular gene expression. An understanding of the mechanism of transformation by these viruses will require at the outset a knowledge of the number and nature of gene functions in the early region of the viral genome. Ultimately, more complex questions concerning alterations in the pattern of expression of the cellular genome will have to be addressed.

The first questions concerning the virus have been approached through the isolation and characterization of conditional lethal virus mutants. Two selective procedures have been used in the case of polyoma virus - temperature-sensitivity and host range. By each of these procedures, non-transforming mutants have been isolated. Temperature-sensitive mutants of the ts-a class are blocked prior to DNA replication in the lytic cycle at the non-permissive temperature. The role of this gene in transformation appears to be one of "initiation" and not "maintenance", since once cells are transformed at the permissive temperature they remain transformed at the non-permissive temperature. These facts con-concerning polyoma virus-cell interactions and genetics of the virus have been extensively reviewed (1,2,3). Host range mutants of the hr-t class do not behave as "early" mutants physiologically, and yet they are blocked in transformation.

In non-permissive mouse cells, they induce the syntheses of T-antigen, viral DNA and viral capsid ~~protein(s)~~ which lead, however, to the production of only 1-5% as much virus as in a wild type infection. Our current knowledge of the polyoma hr-t function and its relation to the ts-a function will be reviewed here. Results are discussed in terms of a model which suggests that the hr-t viral gene is a regulatory gene affecting the pattern of expression of cellular genes.

RESULTS

A. Host Range and Other Properties

Hr-t mutants differ in two important respects from wild type polyoma virus: 1) they have a reduced host range, and 2) they have lost the ability to transform cells. Their essential properties concerning host range are summarized in Tables 1 and 2. Their original isolation relied on polyoma-transformed 3T3 cells as the permissive host, and normal 3T3 cells as the non-permissive host (4). This selection procedure, utilizing the integrated viral genome as a background, would in principle select mutants that are defective in whatever essential functions the integrated virus expresses. Complementation of the mutants by the integrated virus is not necessarily direct, however. This is shown by the fact that cells other than Py-3T3 cells can be permissive for the mutants. Various mouse cells have been screened for their ability to serve as hosts; this is done by infecting parallel cultures with NG-18, prototype of the hr-t class, and wild type virus, and measuring an average burst size for the two viruses. The ratio of burst sizes NG-18 to wild type, called the permissivity value, is typically .01-.05 for 3T3 cells, and .3-.6 for Py-3T3 cells. As shown in Table 2, cells with permissivity values in the range of Py-3T3 exist which do not carry an integrated viral genome. The most interesting examples are the primary cells (early passage mouse embryo fibroblasts and kidney epithelial cells), and murine leukemia virus infected 3T3 cells.

Table 1 - Non-permissive Cell Types ($P \leq 0.1$)

1. 3T3 cell lines (original Swiss, NIH-Swiss, Balb)
2. SV-40 transformed 3T3 cells
3. 3T12 and other spontaneously transformed cell lines
4. 3T3 cells transformed by methylcholanthrene or x-irradiation
5. Multiply passaged but pre-crisis mouse embryo fibroblasts

Table 2 - Permissive Cell Types ($P \geq 0.3$)

1. Polyoma transformed 3T3 cells (original Swiss, NIH-Swiss and Balb)
   a) phenotypic revertants
2. Some C-type RNA viral infected 3T3 cells
   a) MSV-transformed
   b) MLV-infected
3. Certain sub-clones of Balb-3T3 in which MLVs induce morphological and cytopathic change (UC1-B)
4. Primary and early passage mouse embryo fibroblasts, and baby mouse kidney epithelial cells

The ensemble of results on the host range properties have been discussed in terms of a) the existence of cellular function(s) required for lytic virus growth, and b) the ability of the hr-t viral gene to induce the expression of those cellular functions (5,6,7).

A total of 19 mutants have been isolated. They belong to a single complementation group, and all have lost the ability to transform rat or hamster cells (7). The block in transformation is considerably stronger than in lytic growth, the frequency per plaque forming unit being lower by three or more orders of magnitude compared to wild type. NG-18 has also been shown to have lost tumor-inducing ability in newborn hamsters (8). T (tumor) and V (viral) antigens appear in non-permissive mouse cells, and T antigen appears in rat and hamster cells infected by hr-t mutants. Thus, the mutants initiate infections in both kinds of hosts, and are not blocked in the expression of previously recognized early and late viral functions (7).

B. Complementation Between Hr-t and Ts Mutants

Tests for physiological cooperation between hr-t and various ts mutants can be carried out by doubly infecting normal 3T3 cells at 39°C, conditions which are non-permissive for each type of mutant. Such complementation studies have been carried out and the results show that the hr-t group complements with, and is therefore functionally different from, all known classes of ts mutants (9,10). Representative data are shown in Table 3. The efficiency of complementation is given by the complementation factor, i.e., the ratio of the yield in the double infection over the sum of the yields in the single infections.

Table 3 - Complementation between NG-18 and Prototype Mutants of Early and Late Temperature-Sensitive Classes*

Ts parent	Yields of Single Infections		Yields of Mixed Infections	Complementation Factors
	Ts parent	NG-18		
Early:				
Ts-25D	$6x10^2$	$6x10^3$	$3x10^5$	45
Late:				
Ts-10	$1x10^2$	$2x10^4$	$8x10^5$	40
Ts-1260	$2x10^2$	$6x10^3$	$4x10^5$	65

*Infections of NIH-3T3 cells were carried out using multiplicities of 5-20 pfu/cell for the Ts parent and of .1 pfu/cell for NG-18. Incubation was for 60 hours at 39°C, and yields were measured on UC1-B at 33°C.

Complementation between hr-t and ts-a mutants has been studied in detail. It has been shown in each of 18 crosses examined, involving 11 mutants of the hr-t class and 3 mutants of the ts-a class. The complementation is symmetric, i.e., the yields of both mutants increase over the single infection values. Optimal complementation occurs when the ts-a parent is given in 10-50 fold excess over the hr-t parent; this may reflect the need to offset a partial dominant lethal effect of hr-t mutants evidenced by the ability of the latter to depress the yield of wild type in a co-infection (9). Hr-t and ts-a mutants can also co-operate to induce cell transformation in many pair-wise combinations (See Table 4). These results establish the hr-t and ts-a functions as separate, in both the virus growth cycle and in cell transformation (9,10).

Table 4 - Hr-t and Ts-a Mutants Complement for Transformation*

Hr-t	X	Ts-a	Hr-t	Ts-a	Hr-tXTs-a	Complementation Factor
B-2 ($3X10^7$)	X	Ts-48 ($8X10^8$)	0	35	400	11
3A-1 ($1X10^7$)	X	Ts-48 ($5X10^8$)	0	35	360	10
NG-18 ($2X10^6$)	X	Ts-616 ($1X10^7$)	0	1	20	20
NG-18 ($2x10^6$)	X	Ts-a ($1x10^7$)	0	0	10	> 10

Positive complementation was observed in 20 crosses, involving 16 Hr-t class mutants and 4 Ts-a class mutants.

*The results are given as the number of transformants per culture of $5x10^4$ rat or hamster cells, scored as macroscopic clones in soft agar (0.5mm diameter) 15-20 days after infection. Numbers in parentheses are plaque forming units.

C. Mapping by Marker Rescue

As outlined above, the physiological properties of hr-t mutants do not fit the expectations of a typical "early" function, and mutants of this class complement well with both early and late ts mutants. Physical mapping of hr-t mutations on the viral genome was therefore undertaken using the marker rescue technique with restriction enzyme fragments of wild type DNA (11). Figure 1 shows a diagram of the polyoma genome redrawn from the work of Griffin et al., (12) and Kamen et al., (13). The fragments produced by restriction enzymes Hpa-II and Hind-III are shown along with the transcription patterns defining early and late regions. Table 5 shows the results with 3 hr-t mutants and single representatives of various ts mutant groups. The hr-t mutants are rescued by fragment 4, from the proximal (5') part of the early region. Mutants 3A-1 and NG-59 have been further mapped to a Hae-III subfragment of about 350 base pairs coming from the 5' end of Hpa-II fragment 4 (11). Deletions of about 150 base pairs in fragment 4 are seen in NG-18 and some but not all of the other hr-t mutants (7,11).

Figure 1

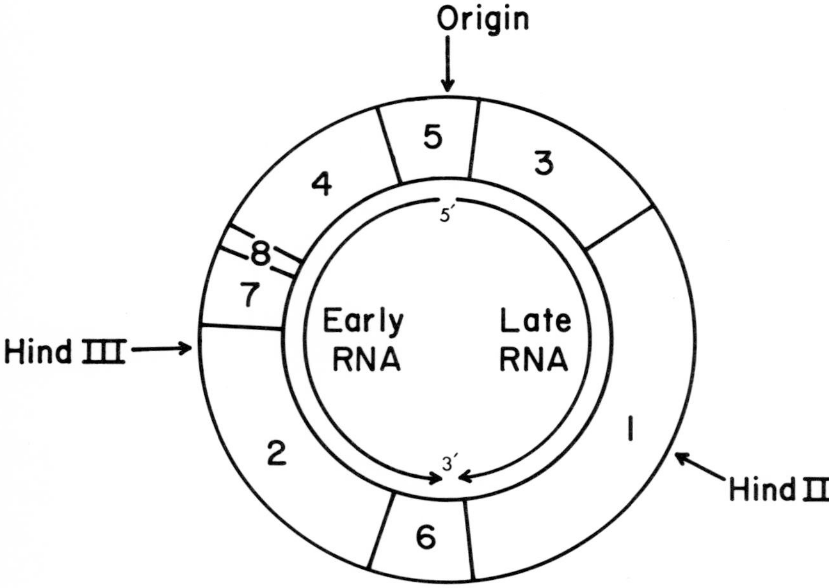

Ts-25D, a ts-a class mutant, is rescued by fragment 2 from the distal (3') part of the early region, and ts-1260, a late mutant is rescued by fragment 1 from the distal part of the late region. The results in Table 5 with these two classes of mutants confirm earlier work by Miller and Fried who mapped several mutants of the ts-a and late classes (14). In addition, they show that ts-3, the non complementing mutant described by Eckhart and Dulbecco (15), is rescued by fragment 3 from the proximal part of the late region.

Table 5 - Rescue of Hr-t and Ts Mutant by Wild-Type HpaII Fragments*

	HpaII fragments							
	1	2	3	4	5	6	7	8
Hr-t class:								
NG-18	1	0	2	35	0	0	0	0
NG-59	0	3	10	160	0	0	0	0
3A-1	5	0	0	50	0	0	0	0
Early class:								
Ts-25D	7	160	0	0	0	0	0	0
Late classes:								
Ts-1260	25	0	0	0	0	0	0	0
Ts-3	1	0	45	0	0	3	0	0

*Results are given as pfu/0.1 ml culture fluid (see ref. 11)

The results on mapping show that the two complementing classes of non-transforming polyoma mutants reside in non-overlapping portions of the early region. Hind-III cleavage of polyoma DNA separates the regions in which ts-a and hr-t mutants map (11,14) (See Figure 1). By ligation of appropriate Hind-III fragments from Ts-25D and NG-18 it has been possible to construct both a wild type virus (56% fragment of Ts-25D + 44% fragment of NG-18) and a double mutant (56% fragment of NG-18 + 44% fragment of Ts-25D). This is shown in Table 6.

The ts-a or distal portion of the early region encodes at least a portion of the T antigen, since ts-a mutants produce a thermolabile antigen (16). Hr-t mutants produce immuno-reactive T antigen in non-permissive cells. The question remains open as to whether the hr-t portion of the early region codes for a protein distinct from the T-antigen.

Table 6 - Ligation of Hind-III Fragments From Hr-t and Ts-a Mutant Viral DNAs*

Virus Source	Perm Cell 32°C	Non-Perm Cell 32°C	Perm Cell 39°C	Non-Perm Cell 39°C
Ligated Recombinants:				
Ts-25D(56%)+NG-18(44%)	280	140	280	280
NG-18(56%)+TS-25D(44%)	120	0	0	0
Virus Controls:				
Wild Type	70	40	40	30
NG-18	110	0	70	0
Ts-25D	200	100	0	0

*Hind-III fragments of DNA from NG-18 and Ts-25D were ligated and used to infect UC1-B cells at 32°C. Plaques resulting from this infection were picked and propagated on UC1-B at 32°C, and assayed as indicated. Results are given as pfu/0.1 ml culture fluid (see ref. 11).

DISCUSSION

Figure 2 shows a model for the action of the hr-t viral gene in both virus growth and cell transformation. The basic proposal of the model is that the hr-t viral gene acts to induce the expression of cellular genes. The action is pleiotropic. Two classes of cellular genes are induced: 1) those giving rise to cellular permissive factors (P) which play a role in the virus growth cycle, and 2) those giving rise to cellular transformation factors (T) which contribute to various aspects of the transformed phenotype. Inductions of these cellular functions by the hr-t viral gene occur not only when the virus is integrated in a transformed cell (as shown in Figure 2), but also during productive infection. Hr-t mutants have lost the ability to induce P and T cellular functions. A full discussion of this model has appeared elsewhere (7,17).

Figure 2

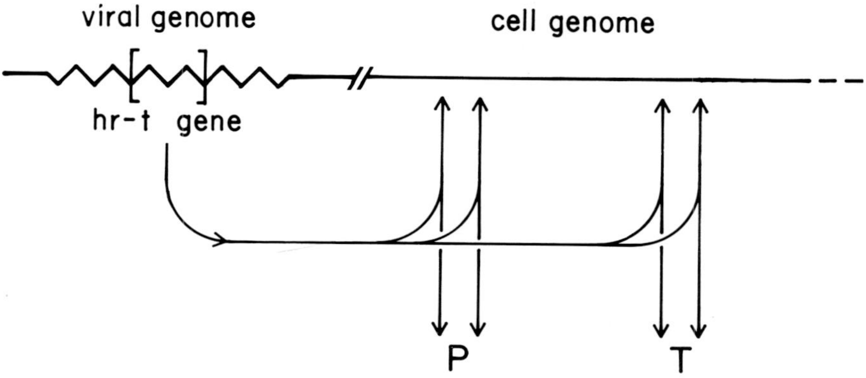

An essential feature of the biology of hr-t mutants is that without exception they express defects in host range and transformation in a co-ordinate fashion. When a normal host range is restored to the mutants by marker rescue, they simultaneously regain wild type transforming ability (11). These observations are consistent with the model, and suggest that the action of the hr-t gene in transformation consists in inducing expression of cellular genes according to the same process which leads to induction of expression of cellular permissive factor(s).

Cell-derived histones present in wild type polyoma and SV-40 virus particles show extensive acetylation of the arginine-rich species H-3 and H-4 compared to the levels seen in normal cellular chromatin. Particles of hr-t mutants have a similar representation of cellular histones, but do not show the high degree of acetylation of H-3 and H-4 (18). These observations imply the induction of a histone acetylation activity based in some manner on the function of the hr-t gene. This can be related to the model in a general way in terms of the putative role of histone acetylation in gene activation (18 and references therein). However, it is not known whether - or to what extent - this activity, clearly evidenced in the virus particles, may affect cellular chromatin; nor is anything known about the process leading from the expression of the hr-t viral gene to histone acetylation.

Figure 3 summarizes some of the important properties of ts-a and hr-t mutants. The properties are organized and

Figure 3

FUNCTIONAL DELINEATION OF EARLY REGION OF POLYOMA VIRAL GENOME

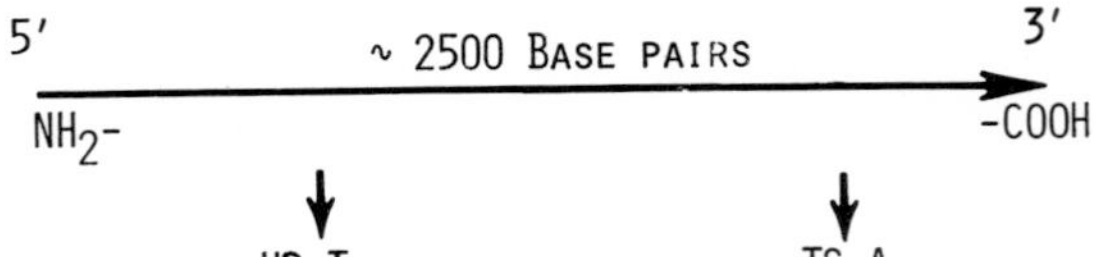

GENE FUNCTIONS:	HR-T	TS-A
PROPERTIES		
A) VIRAL		
T ANTIGEN	+	−
VIRAL DNA	+	−
VIRAL CAPSID	+	−
B) CELL		
ABORTIVE TRANSFORMATION	−	+
CELL SURFACE (CON A)	−	+
HISTONE MODIFICATIONS	−	(+)
OTHER (WSR, TI, MIGRATION, ACTIN)	?	(+)

presented along lines which serve to contrast the mutants in the most obvious way, namely with respect to their abilities to carry out syntheses of viral components and virus-coded products on the one hand, and on their abilities to modify cellular properties on the other. Ts-a mutants behave like typical early virus mutants in the sense that they are altered in a protein (T antigen) made early after infection and required for viral DNA synthesis. Though negative for T antigen, viral DNA and viral capsid, these mutants retain the ability to modify cellular properties under non-permissive conditions. They induce cellular modifications related to the transformed phenotype, as seen most clearly by their ability to induce abortive transformation(19); they also retain wild type ability to modify the cell surface as detected by lectins (20), to decrease wound serum requirement (WSR) and topoinhibition (TI) of cellular DNA synthesis, and to stimulate migration in low serum (21). In contrast, hr-t mutants appear to retain normal or near-normal ability to make known viral products and components, while they have lost the ability to induce at least some of the cellular changes which ts-a mutants still induce. In particular, they are unable to induce abortive transformation assayed as the transient loss of anchorage dependence of growth (22), or to bring about the cell surface change measured by lectins (23). The ability to induce cellular DNA synthesis appears to be retained, somewhat surprisingly, by both types of non-transforming mutant.

The work of this laboratory has been supported by NCI Grant CA16252 and NCI Contract NO1-CP-43299. The following post-doctoral fellows have been involved in the various aspects of the work reported here: Emanual Goldman, Michele Fluck, Roberto Staneloni, Lauren Sompayrac, Jean Feunteun, and Brian Schaffhausen.

REFERENCES

(1) W. Eckhart, Ann. Rev. Genet., (1974) 9: 301.

(2) T. L. Benjamin, Current Topics in Microbiology and Immunology, Springer Verlag, Publ. (1972), 59: 107.

(3) J. Tooze, ed. The Molecular Biology of Tumour Viruses, Cold Spring Harbor Laboratory (1973).

(4) T. L. Benjamin, Proc Nat Acad Sci USA (1970) 67: 929.

(5) T. Benjamin and E. Goldman, Cold Spring Harbor Symposium on Quantitative Biology XXXIX, (1974), p. 41.

(6) E. Goldman and T. L. Benjamin, Virology (1975) 66: 372.

(7) R. J. Staneloni, M. M. Fluck, and T. L. Benjamin, Virology, in press (1977).

(8) R. Siegler and T. Benjamin, Proc Amer Assoc Cancer Research, abstract, (1975), 16:99.

(9) M. M. Fluck, R. J. Staneloni, and T. L. Benjamin, Virology, in press (1977).

(10) Eckhart, W., Virology, in press.

(11) J. Feunteun, L. Sompayrac, M. Fluck, and T. Benjamin, Proc Nat Acad Sci USA, (1976), 73: 4169.

(12) B. E. Griffin, M. Fried, and A. Cowie, Proc Nat Acad Sci USA, (1974), 71: 2077.

(13) R. Kamen, D. M. Lindstrom, H. Shure, and R. W. Old, Cold Spring Harbor Symposium Quantitative Biol. XXXIX, (1974), p. 187.

(14) L. K. Miller and M. Fried, J. Virol.,(1976) 18: 824.

(15) W. Eckhart and R. Dulbecco, Virology (1974) 60: 359.

(16) D. Paulin and F. Cuzin, J. Virol., (1975), 15: 393.

(17) T. L. Benjamin, Cancer Research, (1976) 36: 4289.

(18) B. S. Schaffhausen and T. L. Benjamin, Proc Nat Acad Sci USA, (1976) 73: 1092.

(19) M. Stoker and R. Dulbecco, Nature (London), (1969), 223: 397.

(20) W. Eckhart, R. Dulbecco, and M. M. Burger, Proc Nat Acad Sci USA, (1971) 68: 283.

(21) R. Dulbecco, Proc Nat Acad Sci USA (1970), 67: 1214.

(22) T. Benjamin and L. Norkin, 25th Amer. Symp. on Fund-a mental Cancer Research, M.D. Anderson Hospital and Tumor Institute,(1972), p. 158.

(23) T. L. Benjamin and M. M. Burger, Proc Nat Acad Sci USA, (1970), 67: 929.

Discussion

L. Miller, University of Idaho: Several years ago there was an observation by Vogt and Dulbecco that when large plaque polyoma virus is multiply passaged through transformed cells small plaque polyoma virus is obtained. Mike Fried and I have looked into the mapping of these small plaque viruses from multiple passage through transformed cell lines and found that, in fact, the small plaque morphology was mapped in the late region, -the same region we have mapped as VP1. In doing the ligation of wild type Hind III 55% fragment with the NG-18 Hind III 45% fragment we found that the NG-18 was also carrying a small plaque morphology mutation or variation. I am just pointing out that there may be multiple differences between NG-18 and large plaque Pasadena strains.

T. Benjamin, Harvard Medical School: Lois Miller has pointed out that the small plaque phenotype seems to arise in NG-18. One has to consider various ways in which a small plaque phenotype might arise. It is a rather mutable property, and I think in the case of SV40 also, one can obtain a small plaque virus by changes in and around the origin of DNA replication so that there are various ways to get the small plaque phenotype.

P. Sarin, National Institutes of Health: The P values for your NG-18, whether you use the Py-3T3 or the MLV-3T3 remain the same. I wonder whether the biological and biochemical properties of the two NG-18 are different when grown on either of these two cells?

T. Benjamin: We have not detected differences in any properties depending on the host.

J. Davison, Institute of Cellular Pathology, Brussels: When you began these experiments it was because you expected complementation from the resident polyoma genome in the Py-3T3 cells. I was wondering whether that genome complements tsA mutants of polyoma?

T. Benjamin: Py-3T3 cells do not complement tsA mutants.

THE MOLECULAR BASIS OF TRANSFORMATION BY SIMIAN VIRUS 40

ROBERT G. MARTIN, MARIA PERSICO-DiLAURO
CAROL A.F. EDWARDS and ARIELLA OPPENHEIM
Laboratory of Molecular Biology
NIAMDD
National Institutes of Health
Bethesda, Maryland 20014

Abstract: A model for the molecular basis of transformation which assumes that the T-antigen of SV40 is the only viral product required for transformation and that the T-antigen is an initiator of viral DNA synthesis has been presented. The model proposes that: (i) the T-antigen acting at the end of G_1 drives cells into DNA synthesis under suboptimal growth conditions; (ii) the host in part regulates the expression of T-antigen; (iii) the T-antigen initiates DNA synthesis in the host either by virtue of its high affinity or low affinity binding to DNA; and, (iv) membrane changes are largely secondary to the inability of transformed cells to enter the resting state.

In their original proposal of the replicon model, Jacob, Brenner and Cuzin (1) postulated that DNA synthesis is initiated by a protein having three properties. This initiator protein should act as a positive effector to start DNA synthesis; it should recognize a specific sequence of DNA; and, it should have an affinity for cellular membranes. We now review our hypothesis (2,3) that all of the phenotypic properties of SV40-transformed cells can be accounted for by the action of T-antigen, if: (i) the T-antigen of SV40 is an initiator of DNA synthesis; and, (ii) the expression of T-antigen may be influenced by the host.

T-ANTIGEN IS REQUIRED FOR THE MAINTENANCE OF TRANSFORMATION BY SV40

During transforming infection by SV40, the incoming virus recombines with host DNA and thus becomes stably integrated into the host genome (4,5). The integrated viral DNA can be transcribed into early, _A_-gene mRNA (6-9) which is translated into the T-antigen (10-13). There is much evidence that a functional _A_-gene is required for the maintenance of transformation (14-20). Cells cannot be transformed by viral mutants that contain deletions or substitutions in the _A_-gene (21). Cells transformed by viral mutants that have a temperature-sensitive _A_-gene product, the T-antigen, in general can neither overgrow a normal monolayer nor clone in soft agar at the restrictive temperature. In one exceptional case, a cell line transformed by a _tsA_ mutant could overgrow a normal monolayer but could not clone in soft agar at the restrictive temperature (22). We found that the SV40 T-antigen was produced in excessive amounts in this line and we postulated that the deviant behavior results from 'leakiness' of the temperature-sensitive mutation (22). All of the data are therefore consistent with the idea that a functional _A_-gene is required for the maintenance of transformation.

T-ANTIGEN MAY BE THE ONLY VIRAL PROTEIN REQUIRED FOR TRANSFORMATION BY SV40

Several lines of evidence suggest that the T-antigen is the only viral protein necessary for the maintenance of transformation. (i) Extensive analyses of temperature-sensitive mutants and of deletion mutants have failed to demonstrate the existence of any other early cistron required for SV40 viral replication (2,23,24). (ii) Only the early region is required for transformation (25). (iii) Estimates of the size of the SV40 genome and of the T-antigen suggest that there is insufficient genetic information to code for an additional early protein (19,26). (iv) Mutants of SV40, in which part of the early region proximal to the _A_-gene has been deleted, produce T-antigen of normal molecular weight, are viable and can transform (27,28). In this respect SV40 may differ from the closely related papovavirus, polyoma. HR-T mutations of polyoma, which map in the early region corresponding to the location of the nondefective early deletions of SV40, reduce the frequency of transformation (29). This result might indicate that the polyoma genome encodes a second protein required for transformation. However, it is also possible that the HR-T mutations are altered in a control region or 'leader' sequence required for expression of the polyoma _A_-gene.

Thus, although there is no conclusive proof that the only SV40-encoded protein required for transformation is the T-antigen, there is suggestive evidence in favor of such an hypothesis. We have therefore constructed a model for the molecular basis of transformation which proposes that all of the phenotypic properties of SV40-transformed cells result from the action of T-antigen encoded by the integrated SV40 genome (see below).

T-ANTIGEN IS AN INITIATOR OF DNA SYNTHESIS

The T-antigen of SV40 fulfills the first criterion for an initiator of DNA synthesis (1) by acting as a positive effector to initiate DNA synthesis during lytic infection (24, 30). A functional _A_-gene is required for the initiation of viral DNA synthesis and for the induction of host DNA synthesis in permissive cells (31). Viral DNA replication stops within 20 minutes after a shift to the restrictive temperature in _tsA_-infected monkey cells. And viral DNA synthesis during that interval consists primarily of the completion of replicative intermediates initiated prior to the temperature shift (24,30). T-antigen acts in _trans_ as demonstrated by genetic complementation of _tsA_ mutants by other _ts_ mutants and by wild-type virus (30,32).

The T-antigen is a site-specific DNA-binding protein and thus in part fulfills the second criterion for an initiator protein. Jessel _et al_. (33) have demonstrated that the SV40 T-antigen binds specifically to certain restriction fragments of SV40 with a dissociation constant of the order of 10^{-12} M and nonspecifically to other fragments with a dissociation constant of the order of 10^{-10} M. Curiously, the specific binding sites include at least one site which is distant from the origin of DNA replication.

Indirect evidence suggests that the T-antigen has affinity for membranes, fulfilling the third criterion for an initiator protein (1). SV40-transformed cells contain a new antigen on their membrane surface, which can be recognized by the immune system (34-41). This tumor specific transplantation antigen, TSTA, is virus-specific. We have recently demonstrated that: (i) TSTA appears at the same time as T-antigen during lytic infection (42); (ii) cells transformed by _tsA_ mutants that retain T-antigen at the nonpermissive temperature also retain TSTA, and those that lose T-antigen, lose TSTA (43,44); and, (iii) more TSTA is found in the nucleus than associated with plasma membranes and the intranuclear TSTA copurifies with T-antigen through several steps of purification (45). We have therefore postulated that T-antigen contains TSTA activity and that the TSTA associated with membranes is identical with, or a product of, the T-antigen (45).

Deppert and Walter (46) have shown that a nondefective SV40-adenovirus 2 hybrid that expresses the SV40 TSTA activity produces two SV40-specific polypeptides of 42 and 56K, both of which are associated with the plasma and nuclear membrane fractions. Further, their preliminary data show that the peptide maps of these polypeptides correspond to portions of the peptide map of the T-antigen. These results, considered in conjunction with the size of the T-antigen and the position of TSTA on the genetic map of SV40 (26,47,48), suggest that the carboxy-terminal portion of the T-antigen has a high affinity for membranes (both nuclear and plasma) and that it serves as the immunogen for TSTA.

A MODEL FOR SV40 TRANSFORMATION

Our model for the molecular basis of transformation by SV40 (2,3) proposes the following: (i) The T-antigen interferes with the cell cycle by driving transformed cells into S phase under suboptimal conditions, which cause normal cells to enter a resting state, G_o. (ii) The expression of the A-gene is controlled at least in part by the host and may be determined by the chromosomal site at which the viral DNA is integrated. (iii) The T-antigen initiates DNA replication at sites not serving as origins of DNA replication in nontransformed cells. (iv) Most membrane changes associated with transformation are normal properties of rapidly growing nontransformed cells and hence are secondary to the inability of transformed cells to enter G_o.

(i) Effect of T-antigen on the cell cycle

The normal cell cycle consists of four parts: G_1, the phase of growth prior to DNA synthesis; S, DNA synthesis; G_2, the phase of growth following DNA synthesis; and, M, mitosis. We have proposed that nontransformed cells enter a resting state by leaving the cell cycle when the medium becomes depleted for an essential growth factor (49). A similar view of the resting state has been proposed by Sander and Pardee (50) and by Baserga, Costlow and Rovera (51). According to our model all cells proceed through G_1 to the restriction point, R, a decision point just prior to S phase. [The idea of a restriction point was originally proposed by Pardee (52), who prefers to place R earlier in G_1.] Under optimal conditions cells proceed through R into S phase. We propose that at R, complex controls, possibly using a biochemical cascade mechanism (53), sense whether conditions are optimal. If conditions are favorable for further growth, these controls stimulate the synthesis of the normal host initiator protein(s) for cellular DNA synthesis. If conditions are suboptimal, the cell leaves the cell cycle and reaches a resting state, G_o. According to our model, the T-antigen encoded by SV40, acting as an initiator of host DNA synthesis, drives cells into S phase even under suboptimal growth conditions.

This first aspect of the model allows several predictions. First, T-antigen should be necessary and sufficient for transformation by SV40, but only if it is present at the end of G_1. If T-antigen were synthesized by an integrated SV40 only transiently during the S phase and if T-antigen were degraded prior to the end of G_1, the host cell would be expected to have none of the phenotypic properties of a transformed cell. [Recent evidence suggests a rapid turnover of T-antigen in permissive and transforming infections (26).] A similar phenotype would be expected if the cell expressed T-antigen only during G_o. Second, the model predicts that if cells have reached G_o, the synthesis of T-antigen will not immediately drive them into S phase. This prediction has been verified in a cell line transformed by a <u>tsA</u> mutant of SV40 (49). This cell line enters G_o at 40° in depleted medium, presumably because the T-antigen is functionally inactive. After these cells are shifted to 33° where they can grow in depleted medium, they return to S phase only after a prolonged delay. A third prediction of the model, which concerns the resting state of nontransformed cells, is that certain functions might be expressed only when the cell is in the resting state. We are currently attempting to identify proteins of this type.

(ii) <u>Control of T-antigen expression by the host</u>

The second feature of our model is that the integrated SV40 is not expressed autonomously. [Hence we have referred to the model as the semiautonomous replicon model (2).] We proposed that SV40 could become integrated at different sites in the host DNA and that the host would influence the expression of the virus depending upon its site of integration. Recent evidence demonstrates that SV40 does not have a specific integration site on its DNA and that it is not integrated at a specific site in the host (54-57).

Integration at various sites in the host genome was originally suggested to explain the existence of the intermediate transformants described by Risser and Pollack (58). After 3T3 cells were infected with SV40 and randomly cloned, some clones behaved neither like normal 3T3 cells nor like fully transformed cells. These intermediate transformants were unable to clone in soft agar or overgrow normal monolayers but did grow slightly faster than normal cells in depleted medium. Clones of these intermediate transformants included cells both with and without T-antigen. Subcloning in the absence of selective pressure always gave colonies which were again mixed with respect to T-antigen expression. We proposed (2,3) that the expression of T-antigen in these cells was cell cycle dependent and that these cells might express T-antigen only during a portion of the cell cycle. Experiments to test this hypothesis are in progress.

Basilico and Zouzias (59) demonstrated more directly the dependence of T-antigen expression on the cell cycle. Following transformation of cells by wild-type SV40, a cell line was selected which is temperature-sensitive with respect to its transformed phenotype. The SV40 in this cell line is not temperature-sensitive. At the high temperature these cells enter a resting state and shut off the synthesis of early SV40 mRNA and T-antigen. Our model could explain this behavior assuming that: (a) the SV40 in this cell line is integrated at a host site which is normally expressed only in G_2; and, (b) the temperature-sensitive cell defect acts to prolong G_1 at the nonpermissive temperature. During growth at the permissive temperature, G_1 would be of normal length and sufficient T-antigen would remain at R to drive the cell into a new S phase. However, when G_1 is prolonged due to the temperature-sensitive mutation of the host, T-antigen would be degraded so that the cell would not be driven into S phase, but rather would enter a normal G_o state. Other explanations would also be compatible with our model.

(iii) T-antigen is an initiator of host DNA synthesis

The third aspect of our semiautonomous replicon model is that the T-antigen initiates DNA synthesis by interacting with nucleotide sequences in the host. This initiating interaction could utilize either high or low affinity binding of T-antigen to DNA. If T-antigen carries out its biochemical function only when bound to a specific nucleotide sequence (i.e. when bound with a dissociation constant of $\sim 10^{-12}$ M), then we must speculate that the nucleotide sequence recognized by the T-antigen is as short as 10 to 12 nucleotides. [A sequence of 10 nucleotides occurs at random with a probability of 1 in 4^{10} (or 1 in 10^6 nucleotides); therefore, statistically such a sequence will occur 10^4 times in any mammalian cell, which has $\sim 10^{10}$ nucleotides.] On the other hand, if the T-antigen can carry out its biochemical function when bound nonspecifically (i.e. when bound with a dissociation constant of $\sim 10^{-10}$ M), we need only speculate that T-antigen must be present in sufficient quantity to initiate host DNA synthesis at nonspecific sites.

Either mode of action for the initiation of host DNA synthesis by the T-antigen would predict that an integrated SV40 in a transformed cell would behave as in Fig. 1.

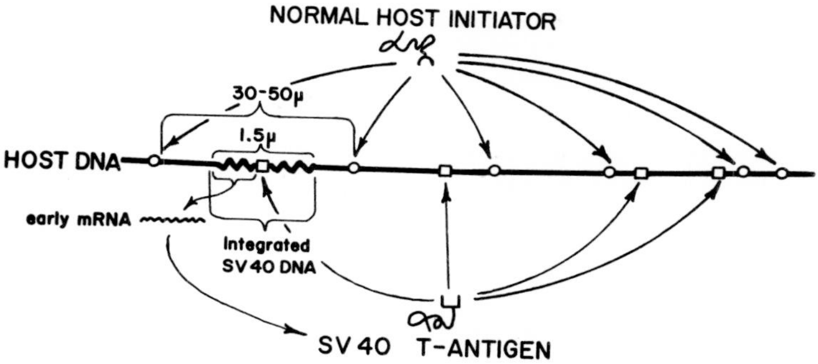

Fig. 1. A stretch of cellular DNA with an integrated viral genome is illustrated. None of the separate structures are drawn to scale. A host initiator protein, whose synthesis is presumed to be strictly regulated and to be limited to the period of time immediately after the restriction point, acts at specific nucleotide sequences represented by the open circles. The average distance between these normal origins of DNA replication is 30-50μ in Chinese hamster ovary cells (60). This distance corresponds to n in equation (1) and is estimatedto be approximately 40 μ in Chinese hamster lung cells (data from Fig. 2.). The integrated SV40 can be transcribed into early mRNA, although this transcription is postulated to be under some host control depending upon the site of SV40 integration. The early mRNA is translated into T-antigen, which acts as an initiator of DNA synthesis. The nucleotide sequences recognized by the T-antigen may be as short as 10-12 bases and therefore could occur by chance in the normal host some 10^4 times per genome equivalent of DNA. There is at least one such sequence in SV40 itself - the origin of viral DNA synthesis. These sequences are represented by the open squares. On the other hand, the open squares could represent nonspecific binding sites. The average distance between these squares corresponds to x in equation (1) and is calculated to be approximately 80-100 μ (data from Fig. 2). Rapidly growing transformed CHL cells seem to initiate DNA synthesis at sites of both types, whereas transformed cells in depleted medium initiate DNA synthesis only at the 'new'

sites. It may be that in the cells of other species, initiation of the 'new' sites precludes initiation at the normal sites.

The T-antigen would interact at sites in the host that are different from the normal sites of host DNA initiation. The relative abundance of the two types of sites cannot be predicted, although estimates of the mean size of vertebrate replicons, 30-60 μ (60-63), would suggest that nontransformed cells contain between 4 and 8 x 10^4 replicons per cell. Different mammalian cells might well have different numbers of sites recognizable by the T-antigen or different amounts of T-antigen so that the frequency of sites at which interaction occurs might differ from hamster to mouse to man. If rapidly growing transformed cells of some species initiate DNA synthesis at the normal sites as well as at the sites activated by T-antigen, the number of origins of DNA replication per cell should increase, and the average replicon size should decrease. A decrease in replicon size need not shorten the length of S phase, because of the complex programming of DNA replication.

Preliminary results suggest that the number of replicons per cell increases following transformation of Chinese hamster lung cells by SV40. Cells growing under optimal conditions were labeled with ^{3}H-thymidine for 8 minutes and the DNA fibers were autoradiographed according to the procedure of Huberman and Riggs (60) as modified by Hand and Tamm (61). In these experiments the replication forks along the DNA fiber are visualized. Several hundred sets of tracks were analyzed for replication forks. The results of experiments with transformed and nontransformed cells are presented as histograms in Fig. 2. Nontransformed CHL cells showed a modal number of approximately 6 replication forks per 130 μ, giving an average replicon size of ∿ 42 μ [assuming 2 forks per replicon since mammalian DNA synthesis is mostly bidirectional (60)]. The same results were obtained for cells grown at 33° or 40°. On the other hand, CHL cells transformed by wild-type SV40 had approximately 9 forks per 130 μ or an average replicon size of ∿ 29 μ. Again, this decrease in replicon size and increase in the number of replicons was independent of temperature. As expected, a cell line transformed by a tsA mutant of SV40 had a nearly normal replicon size of approximately 40 μ when grown at 40°; at this temperature the T-antigen is inactive and the cell does not exhibit the transformed phenotype. In contrast, when grown at 33°, these cells gave an average replicon size of ∿ 30 μ, similar to the replicon size of cells transformed by wild-type SV40; at this temperature the cell line exhibits the transformed phenotype.

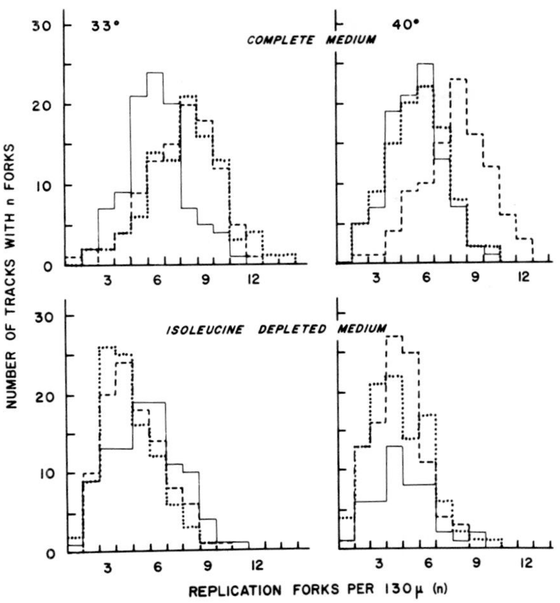

Fig. 2. The number of sets of tracks observed with n forks per set is plotted against the number of forks per 130 μ. Randomly growing normal or transformed CHL cells growing on slides were pulse-labeled with ^{3}H-thymidine (500 μCi/ml; 50 mCi/μmole) and 2 x 10^{-6}M FdUrd after pretreatment for 30 min with FdUrd (2 x 10^{-6}M) as described by Hand and Tamm (61). The cells were lysed and spread. After washing the slides were covered with Kodak Emulsion NTB 3, exposed for 3 months at -15° and developed. The slides were examined at a magnification of 600X in a Leitz binocular microscope equipped with lens markings which correspond to 130 μ at this magnification. In each case the solid lines (——) correspond to non-transformed CHL cells, the dashed lines (---) to cells transformed by wild-type SV40 and the dotted lines (•••) to cells transformed by the SV40 mutant, tsA239. The cells were growing in Dulbecco-Vogt medium with 10% fetal bovine serum (upper panels) or in isoleucine-depleted medium containing 2.5% serum (lower panels) and at 33° (left hand panels) or 40° (right hand panels). In most cases 100 sets of tracks were analyzed under each set of conditions for each cell line.

From these data one can calculate the approximate distance, x, between the new origins of DNA replication activated by the T-antigen. This distance is given by the equation:

$$\frac{1}{x} = \frac{1}{t} - \frac{1}{n} \qquad \text{Eq. (1)}$$

where t = the average replicon size in transformed cells and n = the average replicon size in nontransformed cells. For our experiments, t = 29 μ, n = 42 μ and, therefore, x = 94 μ.

In depleted medium transformed cells would be expected to initiate DNA synthesis only at sites activated by T-antigen. Autoradiograms were prepared using cells incubated for 2 days in depleted medium. Many tracks were found on the slides prepared from the transformed cells, and the average replicon size now appeared to be about 80 μ, in agreement with our calculated value. On the other hand, very few tracks were visualized on the slides prepared from the starved normal CHL cells; these cells appeared to have the same replicon size as rapidly growing CHL cells (Fig. 2). Our observations are consistent with the model (Fig. 1) and suggest that in Chinese hamster lung cells the average distance between the new initiation sites resulting from interaction with the T-antigen is ∿ 80 μ, whereas the distance between normal host sites is ∿ 40 μ. Indeed, previous estimates of the distance between replicons in hamster cells have been 30-50 μ (60,63). Although our experiments strongly suggest that the number of origins for DNA replication increases after transformation by SV40, they cannot, however, determine whether the new initiation sites are different from those employed in normal DNA replication.

(iv) The membrane changes following transformation by SV40 are secondary effects

Our model implies that changes in the cell surface associated with transformation [see Tooze, (64)] are, in fact, not specific for transformation, but are, instead, characteristic of normal growth. Indeed, Weber (65) has concluded that most, if not all, membrane properties associated with transformation are also observed in normal rapidly growing cells.

REFERENCES

(1) F. S. Jacob, S. Brenner and F. Cuzin. Cold Spring Harb. Symp. Quant. Biol. 28 (1963) 329.

(2) R. G. Martin, J. Y. Chou, J. Avila and R. Saral. Cold Spring Harb. Symp. Quant. Biol. 39 (1974) 17.

(3) R. G. Martin, M. Persico-DiLauro, C. A. F. Edwards and A. Oppenheim. Proceedings of the International Union of Biochemistry Varanasi, India (in press) (1976).

(4) J. Sambrook, H. Westphal, R. Srinivasan and R. Dulbecco. Proc. Nat. Acad. Sci. USA 60 (1968) 1288.

(5) L. D. Gelb, D. E. Kohne and M. A. Martin. J. Mol. Biol. 57 (1971) 129.

(6) G. Sauer and J. R. Kidwai. Proc. Nat. Acad. Sci. USA 61 (1968) 1256.

(7) M. A. Martin. Cold Spring Harb. Symp. Quant. Biol. 35 (1970) 833.

(8) J. Sambrook, P. A. Sharp and W. Keller. J. Mol. Biol. 70 (1972) 57.

(9) G. Khoury, M. A. Martin, T. N. H. Lee and D. Nathans. Virology 63 (1975) 263.

(10) P. H. Black, W. P. Rowe, H. C. Turner and R. J. Huebner. Proc. Nat. Acad. Sci. USA 50 (1963) 1148.

(11) J. C. Alwine, S. I. Reed, J. Ferguson and G. R. Stark. Cell 6 (1975) 529.

(12) T. Kuchino and N. Yamaguchi. J. Virol. 15 (1975) 1302.

(13) D. G. Tenen, P. Baygell and D. M. Livingston. Proc. Nat. Acad. Sci. USA 72 (1975) 4351.

(14) J. S. Brugge and J. S. Butel. J. Virol. 15 (1975) 619.

(15) G. Kimura and A. Itagaki. Proc. Nat. Acad. Sci. USA 72 (1975) 673.

(16) R. G. Martin and J. Y. Chou. J. Virol. 15 (1975) 599.

(17) R. G. Martin and J. L. Anderson, in: Biology of Radiation Carcinogenesis, eds. J. M. Yuhas, R. W. Tennant and J. B. Regan (Raven Press, New York, 1976) p. 287.

(18) M. Osborn and K. Weber. J. Virol. 15 (1975) 636.

(19) M. Osborn and K. Weber. Cold Spring Harb. Symp. Quant. Biol. 39 (1974) 267.

(20) P. Tegtmeyer. J. Virol. 15 (1975) 613.

(21) W. A. Scott, W. W. Brockman and D. Nathans. Virology 75 (1976) 319.

(22) D. G. Tenen, R. G. Martin, J. Anderson and D. M. Livingston. J. Virol. (in press) (1977).

(23) J. Y. Chou and R. G. Martin. J. Virol. 13 (1974) 1101.

(24) J. Y. Chou, J. Avila and R. G. Martin. J. Virol. 14 (1974) 116.

(25) P. J. Abrahams, C. Mulder, A. van de Voorde, S. O. Warnaar and A. J. van der Eb. J. Virol. 16 (1975) 818.

(26) P. Tegtmeyer, M. Schwartz, J. K. Collins and K. Rundell. J. Virol. 16 (1975) 168.

(27) T. E. Shenk, J. Carbon and P. Berg. J. Virol. 18 (1976) 664.

(28) K. Rundell, J. K. Collins and P. Tegtmeyer. J. Virol. (in press) (1977).

(29) J. Feunteun, L. Sompayrac, M. Fluck and T. Benjamin. Proc. Nat. Acad. Sci. USA 73 (1976) 4169.

(30) P. Tegtmeyer. J. Virol. 10 (1972) 591.

(31) J. Y. Chou and R. G. Martin. J. Virol. 15 (1975) 145.

(32) J. Y. Chou and R. G. Martin. J. Virol. 15 (1975) 127.

(33) D. Jessel, T. Landau, J. Hudson, T. Lalor, D. Tenen and D. M. Livingston. Cell 8 (1976) 535.

(34) K. Habel and B. E. Eddy. Proc. Soc. Exp. Biol. Med. 113 (1963) 1.

(35) B. E. Eddy, G. E. Grubbs and R. D. Young. Proc. Soc. Exp. Biol. Med. 117 (1964) 575.

(36) H. Goldner, A. J. Girardi, V. M. Larsen and M. R. Hilleman. Proc. Soc. Exp. Biol. Med. 117 (1964) 851.

(37) A. J. Girardi. Proc. Nat. Acad. Sci. USA 54 (1965) 445.

(38) S. S. Tevethia and F. Rapp. Proc. Soc. Exp. Biol. Med. 123 (1966) 612.

(39) J. H. Coggin, L. H. Elrod, K. R. Ambrose and N. G. Anderson. Proc. Soc. Exp. Biol. Med. 132 (1969) 328.

(40) A. J. Girardi and V. Defendi. Virology 42 (1970) 688.

(41) M. S. Drapkin, E. Appella and L. W. Law. J. Natl. Cancer Inst. 52 (1974) 259.

(42) J. L. Anderson, R. G. Martin, C. Chang and P. T. Mora. Virology (in press) (1977).

(43) J. L. Anderson, C. Chang, P. T. Mora and R. G. Martin. J. Virol. (in press) (1977).

(44) C. Chang, J. L. Anderson, R. G. Martin and P. T. Mora. J. Virol. (in press) (1977).

(45) J. L. Anderson, R. G. Martin, C. Chang, P. T. Mora and D. M. Livingston. Virology (in press) (1977).

(46) W. Deppert and G. Walter. Proc. Nat. Acad. Sci. USA 73 (1976) 2505.

(47) A. M. Lewis, Jr. and W. P. Rowe. J. Virol. 12 (1973) 836.

(48) P. Lebowitz, T. J. Kelly, Jr. D. Nathans, T. N. Lee and A. M. Lewis, Jr. Proc. Nat. Acad. Sci. USA 71 (1974) 441.

(49) R. G. Martin and S. Stein. Proc. Nat. Acad. Sci. USA 73 (1976) 1655.

(50) G. Sander and A. B. Pardee.. J. Cell Physiol. 80 (1972) 267.

(51) R. Baserga, M. Costlow and G. Rovera. Fed. Proc. 32 (1973) 2115.

(52) A. B. Pardee. Proc. Nat. Acad. Sci. USA 71 (1974) 1286.

(53) E. R. Stadtman, P. B. Chock and S. P. Adler, in: Israel Scientific Research Conf., Vol. III, Metabolic Interconversion of Enzymes, in press (1977).

(54) C. M. Croce. Proc. Nat. Acad. Sci. USA (in press) (1977).

(55) C. M. Croce, K. Huebner, A. J. Girardi and H. Koprowski. Cold Spring Harb. Symp. Quant. Biol. 39 (1974) 335.

(56) M. Botchan, W. Topp and J. Sambrook. Cell 9 (1976) 269.

(57) G. Ketner and T. J. Kelly, Jr. Proc. Nat. Acad. Sci. USA 73 (1976) 1102.

(58) R. Risser and R. Pollack. Virology 59 (1974) 477.

(59) C. Basilico and D. Zouzias. Proc. Nat. Acad. Sci. USA 73 (1976) 1931.

(60) J. A. Huberman and A. D. Riggs. J. Mol. Biol. 32 (1968) 327.

(61) R. Hand and I. Tamm. J. Mol. Biol. 82 (1974) 175.

(62) H. G. Callan. Cold Spring Harb. Symp. Quant. Biol. 38 (1973) 195.

(63) D. Housman and J. A. Huberman. J. Mol. Biol. 94 (1975) 173.

(64) J. Tooze. The Molecular Biology of Tumor Viruses (Cold Spring Harb. Lab., New York, 1973).

(65) M. Weber. J. Cell Physiol. Supplement (in press) (1977).

Discussion

P. Duesberg, University of California, Berkeley: How do you explain the old findings that there are flat variants which do contain T-antigen, yet do not show the transformed signature?

R. Martin, National Institutes of Health: The story on the flat revertants is a very interesting one and relates precisely to what I talked about, concerning the multiplicity of infection. All of the earlier flat revertants isolated by Bob Pollack were isolated in a cell line - SV101 - which contains many, many copies of SV40. Therefore it is not at all surprising that the only way one can get a flat revertant from such cells is by getting some secondary, down-the-line mutation. Joe Sambrook has gotten a rat cell line which has only a single copy of SV40. Bob Pollack has taken that cell line and isolated eight new flat revertants. It turns out that everyone of these new flat revertants is T-antigen negative.

P. Duesberg: They have only a few copies, don't they - one or two copies? So do you know the concentration of the T-antigen in flat variants? Have you measured that?

R. Martin: No, I haven't.

A. Koch, Indiana University: There is considerable variability in the life length of individual cells within a population of cells in balanced growth. The age at division typically varies 20% no matter what growth condition or types of cells are considered. Recently Smith and Martin in England have suggested that most of this variability in eukaryotic cells arises in G_r at the time that might correspond to the T-antigens by the pass proposed here. Has the distribution of life lengths of transformed cells within a culture been measured? Does it have a smaller coefficient of variation than untransformed cells?

R. Martin: It is a very complicated question, because the growth rate of any cell depends upon what medium you are growing them in. You can change the growth rate of even transformed cells very markedly by growing them in depleted medium or by growing them in pure serum. The difference in the growth rate is largely in the length of time for protein synthesis which occurs largely in G1, and therefore one preferentially prolongs the length of G1. The highest rate is the same in transformed and normal cells. You cannot talk about the lowest rate because normal cells will go into a resting state and therefore give you an infinite generation time.

A. Koch: For most eukaryotic cells, there is a variability of about 20 percent in the life length of individual cells. Tremendous variability in cells growing in the same population comparing cells that just happen to divide fast and those that happen to divide slow. There is theory in the literature, suggested by several people, that most of this variability arises by variability in G1 alone and so, possibly your transformed cells would have gotten around a variable time and the individual cells in your transformed clone might behave much more one like the other than individual cells in an untransformed clone.

R. Martin: I have not looked at the rate of growth of individual cells in a population, which is what you would have to do.

D. Billen, Oak Ridge National Laboratory: Do you have any information as to whether the T-antigen is necessary once you have triggered off the S-phase (G1 $\rightarrow$ S), for completion of S-phase, i.e. if production was terminated by a shift to a non-permissive temperature, would S-phase be completed?

R. Martin: I have no information on that, other than in lytic infection of permissive cells.

S. O'Brien, National Institutes of Health: Implicit in your interpretation, and also that presented by Dr. Livingston this morning, was the participation of the T-antigen somehow in either the DNA synthesis or the transcription of the virus itself, as well as the host as you were discussing. I was wondering if you have any information, or maybe Dr. Livingston does, on the actual effect of T-antigen on DNA synthesis or transcription of the virus in vitro in a reconstruction type experiment ?

R. Martin: I am not sure I understand the question. If you mean, do we have an in vitro system in which T-antigen initiates DNA synthesis, the answer is certainly not. No one has any system in which initiation of DNA synthesis occurs.

INTERACTION OF SV40 T ANTIGEN WITH SELECTED REGIONS OF SV40 DNA

D. M. Livingston, D. G. Tenen, D. Jessel, L. L. Haines, V. Woodard, A. P. Modest, A. Maxam*, and J. Hudson.

The Sidney Farber Cancer Institute and the Departments of Medicine, Peter Bent Brigham Hospital and Harvard Medical School, Boston, MA.; and the *Department of Biochemistry and Molecular Biology, Harvard University, Cambridge, MA.

Abstract: SV40 T antigen is a virus-encoded protein which is the product of the viral A gene. As such, it is required for the initiation of viral DNA replication in infected permissive cells and for the maintenance of transformation in SV40 transformed cells. It is a DNA binding protein. The protein has been extensively purified to date and can be shown to have the capacity to bind to specific regions of the viral genome with high affinity. One of these contains the origin of viral DNA replication. The others lie within the early region. While the significance of these phenomena is not known at present, certain testable possibilities are suggested.

INTRODUCTION

SV40 is a small DNA virus capable of transforming cells from a variety of animals. Its genome is a circular, double-stranded DNA molecule of approximately 3.6×10^6 daltons. Complementation studies with a large number of temperature sensitive mutants reveal the existence of at least three complementation groups all of which have been mapped (1,2). One of these complementation groups, cistron A, gives rise to a protein product which is synthesized prior to the onset of viral DNA replication and is apparently required for both the initiation of viral DNA replication and the maintenance of SV40 induced neoplastic transformation (1-6). The other two cistrons are expressed late in the infectious cycle. Their products have also been identified and are the virion proteins VP_1, VP_2, and VP_3 (7-9). None of these proteins is believed to play any role in the processes of viral DNA replication initiation or transformation.

It has been known for several years that virtually all SV40 transformed cells synthesize a nuclear antigen called (T)umor antigen (10,11). This protein was also recognized several years ago as being synthesized in infected permissive cells prior to the onset of viral DNA replication (12). It is, therefore, an early protein. Abundant evidence accumulated recently indicates that T antigen is a product of viral cistron A (13-19). As such it is strongly suspected of playing a major role in the two aforementioned processes for which temperature sensitive mutant studies have implicated a role for the gene A product.

One of our major goals is to begin to understand how T antigen functions biochemically in these two events. It is known to be a DNA binding protein, with a selective affinity for double-stranded DNA (20). In this regard, we have attempted to determine whether T antigen is capable of interacting with the SV40 genome in a site-specific manner. In particular, evidence will be presented which indicates that T antigen can bind to selected regions of viral DNA, presumably as a result of its relatively selective affinity for certain DNA sequences (21). Reed and co-workers have observed, independently, that this protein can bind to a region of the DNA which is close to or at the origin of viral DNA replication (22).

RESULTS

Purification of SV40 T Protein: We have now found it possible to resolve T antigen from protein contaminants to a high degree. The SV40 transformed human cell line, SV80, is routinely used in our laboratory as a source of the antigen, since it is known to be a super-producer of this protein (23). When purified nuclei from mass cultures of this line are extracted and the antigen purified by a series of steps including ammonium sulfate fractionation, DEAE cellulose chromatography, heparin-Sepharose chromatography, and glycerol gradient sedimentation, a fraction is obtained which, following radioiodination, appears during SDS gel electrophoresis as a species of approximately 90,000 molecular weight (Fig.1).

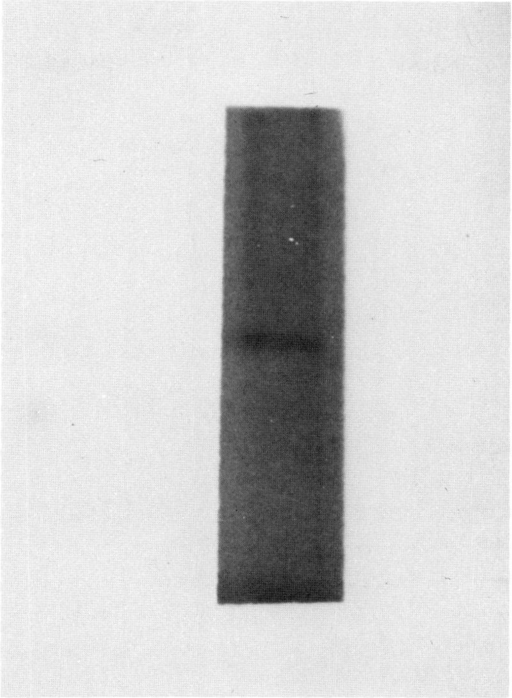

Fig. 1. SDS Gel Electrophoresis of ^{125}I-Glycerol Gradient Fraction. The glycerol gradient fraction noted in Table 1 was iodinated by the method of Hunter and Greenwood (36). Following separation of the labelled protein from free Na ^{125}I, the former was analyzed by electrophoresis in a 10% polyacrylamide slab gel. Gel electrophoresis conditions were those described by Laemmli (37).

We have observed the migration of this band in gels of various porosity, run in different buffers to be somewhat anomalous in that the values obtained for molecular weight are not constant. The observed range is between 86,000 and 96,000 daltons.

The radioiodinated band noted in Fig. 1 carries one or more T antigenic determinants. As shown in Fig. 2, it binds selectively to a hamster anti-T IgG Sepharose immunoadsorbent but not to a column of the same size containing identical quantities of non-immune IgG.

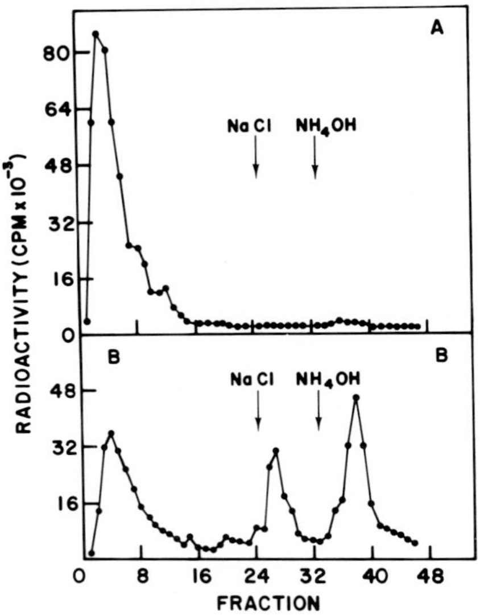

Fig. 2. Immunoaffinity Chromatography of ^{125}I-Glycerol Gradient Fraction. Identical aliquots of the fraction noted in Fig. 1 were applied to two columns: (A) a non-immune hamster IgG Sepharose column and (B) a hamster anti-T IgG Sepharose column, equilibrated in 0.05M Tris·HCl, pH 7.8, 10^{-3}M dithiothreitol (DTT), 20% glycerol, 0.2% Triton X-100, 1 mg/ml bovine serum albumin. After sample application, both columns were washed with 0.05M Tris·HCl, pH 7.8, 10^{-3}M DTT, and then eluted successively with 0.05M Tris·HCl, pH 7.8, 0.10M NaCl, 10^{-3}M DTT and finally with 0.05M NH_4OH. Aliquots of each fraction were counted in Aquasol scintillation

fluid. Each column volume was 3 ml. The volume of each fraction was 2 ml. The IgG employed in the CNBr coupling to Sepharose in each instance had been purified from serum by $(NH_4)_2SO_4$ fractionation followed by DEAE cellulose chromatography. The protein content of both adsorbents was approximately 6 mg/gram Sepharose. All chromatographic operations were at 4°C.

Material eluted from such a column both in NaCl and in the presence of ammonium hydroxide migrated identically on a gel with the starting material (Fig. 3).

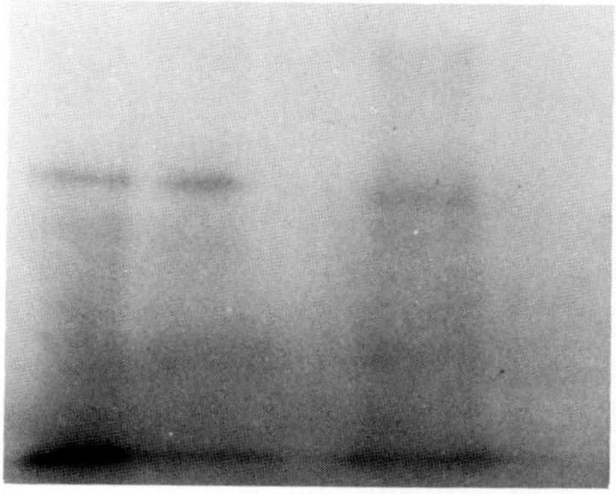

Fig. 3. SDS Gel Electrophoresis of ^{125}I-Glycerol Gradient Fraction Protein Before and After Immunoaffinity Chromatography. Material from both the NaCl and NH_4OH eluted peaks in Fig. 2B was dialyzed against 0.001M Tris, pH 7.2, and then lyophilized and the dried material redissolved in 0.03 ml of Laemmli (37) sample buffer and electrophoresed in parallel with an aliquot of the ^{125}I-glycerol gradient fraction noted in Fig. 1. A 10% polyacrylamide slab gel was employed and the electrophoretic conditions were those described by Laemmli (37). From left to right the lanes are ^{125}I-glycerol gradient fraction, NaCl eluted fraction, NH_4OH eluted fraction.

Moreover, the ^{125}I band can be immunoprecipitated with hamster anti-T IgG serum and second antibody and not with non-immune hamster IgG and second antibody (data not shown). As much as 60% of the input counts can be precipitated under these conditions. The antigen is 200-fold purified at this point and is recovered in approximately 10% yield (Table 1). Experiments are now in progress to determine whether, in fact, this material is still modestly impure or whether it is substantially enough resolved to proceed along with additional protein chemical experimentation.

TABLE 1

Purification of SV40 T Antigen

Fraction	Total CF Units [a]	Protein Concentration (mg/ml)	Specific Activity
Nuclear Extract	2600	16	1.3
20-60% $(NH_4)_2SO_4$	1000	18	2.8
DEAE Pool	480	1	4
Heparin Sepharose Pool	320	0.4	25
Glycerol Gradient Pool	300	0.05	300

[a] One complement fixation (CF) unit represents a 10% deflection of A^{413} employing a standard micro-CF assay in which 1.2 units of complement and 2 units of hamster anti-T antibody are employed in a 0.9 ml reaction mixture (24).

Interaction of SV40 T Antigen with Specific Regions of the SV40 Genome: We have elected to begin to study the nature of the T antigen DNA binding property. In particular, we have chosen to evaluate whether T antigen can bind to the SV40 genome in a site specific manner.

First, a standard nitrocellulose filter assay which serves to trap protein-dsDNA complexes on filters was found to be useful for measuring binding of T antigen to various DNA molecules (24). As shown in Fig. 4, T antigen from the same preparation described in the legend to Fig. 1 can bind virtually all of the molecules in a fixed aliquot of radioactive SV40 DNA I.

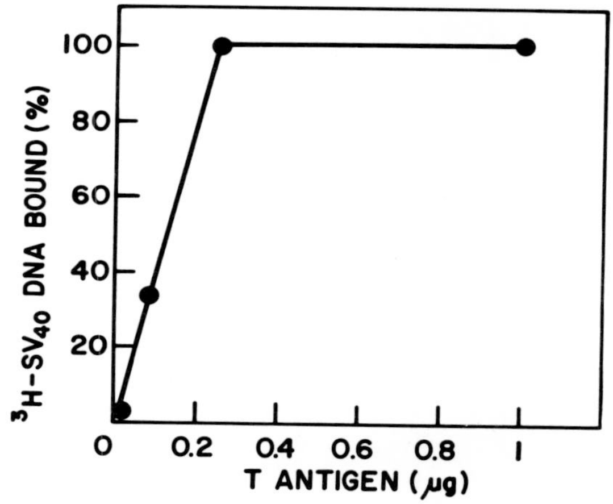

Fig. 4. Binding of ^{3}H-SV40 DNA I by Glycerol Gradient Purified T Antigen. The glycerol gradient fraction noted in Table 1 was employed here. The specific activity of the DNA was 2.2 x 10^5 cpm/μg. The DNA was prepared and binding was measured as described previously (21). 440 cpm of radioactive DNA were present in each 0.25 ml reaction mixture. Following a 10 minute incubation at room temperature, 0.20 ml were filtered.

It was also observed that the antigen can interact with a variety of superhelical DNAs including ØX RF1, λ supercoils, and polyoma DNA I (24). Thus, under at least one set of conditions, the protein is capable of binding non-specifically to a variety of DNA molecules.

T antigen binds equally well, under some conditions, to supercoiled and full-length, linear SV40 DNA (21). Thus, it is possible to readily examine the question of whether T binds selectively to one or more specific regions of the viral genome. To test this, we have fragmented radioactive SV40 DNA with a series of bacterial restriction endonucleases, isolated sets of specific DNA fragments and evaluated the relative ability of T antigen to bind to each. As noted in the physical map depicted in Fig. 5, four sets of fragments have been examined - the Eco R_I + Hpa II 74% and 26% fragments, the four Hpa I + Eco R_I fragments, the six Hin dIII fragments, and the eleven Hin d(II + III) fragments.

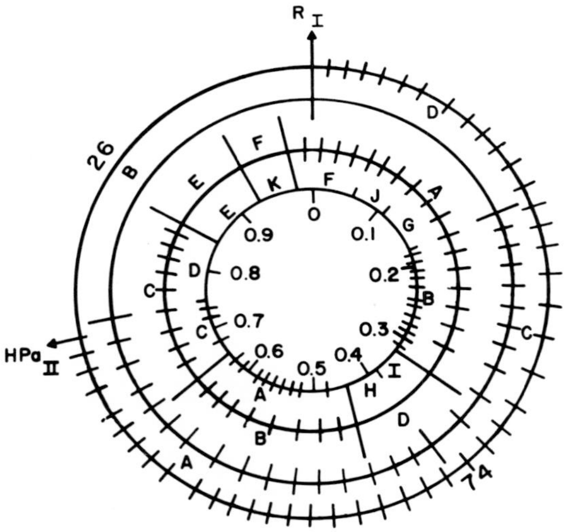

Fig. 5. Physical Map of SV40 DNA. Shown here are four restriction enzyme cleavage maps - from the outside in: cleavage by Eco R_I and Hpa II; Eco R_I and Hpa I; Hin dIII; Hin d(II + III). The hatched areas represent fragments preferentially bound by SV40 T antigen. These include: Eco R_I/Hpa II 76% fragment; Hpa I/Eco R_I A and C; Hin dIII A, B and C; Hin d(II + III) A, B and C.

A typical result is shown in Fig. 6, in which high specific activity Hin dIII fragments exogenously labelled with ^{32}P by the "nick translation" method (25) have been studied.

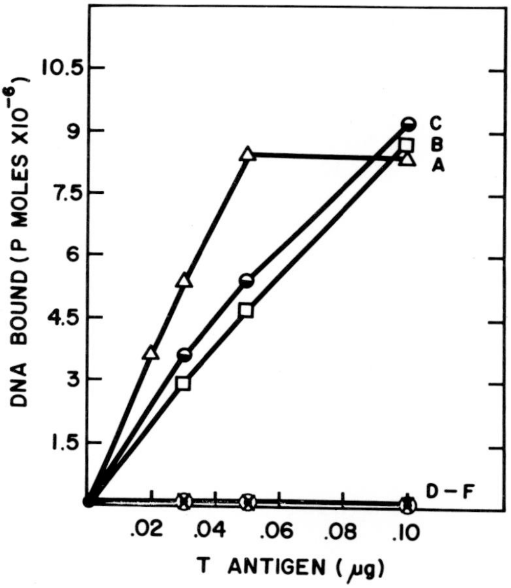

Fig. 6. Binding of ^{32}P Hin dIII Fragments by Partially Purified SV40 T Antigen. The binding assay was performed as noted in the legend to Fig. 4. The T antigen employed here was purified by $(NH_4)_2SO_4$ fractionation, DEAE cellulose chromatography and Agarose A 1.5 m chromatography as noted previously (24). The DNA fragments were purified by digesting non-radioactive SV40 DNA I with Hin dIII, "nicked translating" the restriction digest in the presence of four α-^{32}P deoxynucleoside triphosphates (25), and then separating the fragments by polyacrylamide gel electrophoresis. Fragments, identified by autoradiography, were then eluted from crushed gel slices by soaking. They were subsequently precipitated with ethanol and redissolved prior to use in this and other studies. The specific activity of the DNA used here was 2 x 10^7 cpm/μg. All DNA fragments were present in individual reaction mixtures at a concentration of 2 x 10^{-13}M.

Fragments A, B, and C bind, while D-F do not under these conditions. Also noted in Fig. 5 in the hatched areas, are fragments from each set which bind to T antigen preferentially. As can be seen, the results of all four fragment sets are internally consistent. Preferred T antigen binding sites can be identified in the regions identified by Hin d(II + III) fragments A, B, and C.

These results do not reflect a preference by T antigen for fragments solely on the basis of their relative size, for several reasons. Although Hpa I B is significantly larger than Hpa I C, it binds significantly less well than the latter. Hin d(II + III) C binds in preference to Hin d(II + III) D and yet the two only differ in length by approximately 25-50 base pairs out of 525 (26). Furthermore, we have simultaneously examined the comparative binding of Hin dIII A, B, and C, Hin d(II + III) A and B, and the Eco R_I + Hpa II 26% fragment (21). All are approximately 1000 base pairs in length or greater, yet the Eco R_I + Hpa II 26% fragment binds with significantly lower affinity than all of the former five fragments (21). We have also shown that these results are not specific to the interaction of T antigen with DNA from a particular strain of virus. Moreover, it is likely that the DNA is binding to T antigen, itself, as opposed to a complex of T with another protein since, as shown in Fig. 7, the material represented in Fig. 1 also binds Hin dIII A, B, and C in preference to D-F.

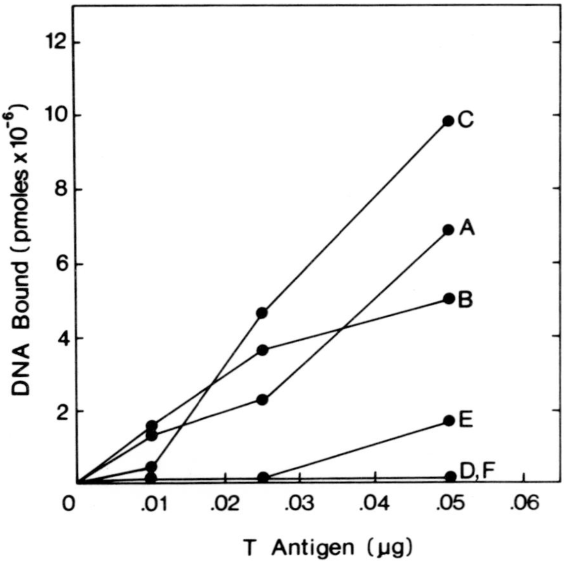

Fig. 7. Interaction of Glycerol Gradient Purified T Antigen with Various Hin dIII Fragments. Aliquots of glycerol gradient T antigen described in Table 1 were mixed with equimolar (3×10^{-13}M) amounts of each of a series of ^{32}P-SV40 Hin dIII DNA fragments by the published method (21). The specific activity of the DNA was 1.3×10^{7} cpm/µg.

The affinity of the protein for these purified binding regions indicated above has been studied. As shown in Table 2, T antigen binds with high affinity to the three Hin dIII fragments suggested to contain preferred binding sites. The dissociation constant (K_D) for each is of the order of 10^{-12}M.

TABLE 2

Affinity of SV40 T Antigen for Hin dIII A-C

Hin dIII Fragment	$[DNA]_{1/2}$	$[T\ Ag]_o$	K_D (M x 10^{-12}M)
A	3.8	5.4	1.1
B	2.7	1.9	1.75
C	2.5	3.3	0.85

The equilibrium dissociation constants (K_D) were calculated as suggested by Riggs, et al. (38), using the equation $[DNA]_{1/2} = K_D + {1/2}[T\ Ag]_o$.

All of the above studies were performed with T antigen isolated from one line of SV40 transformed human cells. The binding of T antigen from SV40 transformed Chinese hamster lung cells has also been evaluated, and similar results have been obtained (21). Thus, the binding phenomena are not specific to T antigen from a given SV40 transformed cell line. We have not yet studied the binding of T antigen isolated from SV40 infected green monkey cells, thus we do not yet know whether "lytic" T antigen has similar DNA binding properties.

Thus, T antigen from more than one species of SV40 transformed cell has the ability to bind to SV40 DNA in at least three discrete regions. While these observations, of themselves, cannot be immediately considered of physiologic significance, they do suggest certain hypotheses. Hin d(II + III) C contains the origin of SV40 DNA replication (27,28). Electron microscopic studies have already suggested that T may interact with one or more sites close to or at the origin of DNA replication (22). T is a product of gene A and as such plays a necessary role in the initiation of viral DNA replication (1,13-17,29). Thus, binding in this region may be hypothesized to be a partial reflection of a role of T in this process - i.e., interaction with specific DNA sequences leading toward the commencement of viral DNA synthesis.

The exact sites of T binding within Hin d(II + III) A and B are not fully known at present but are under investigation. However, one or both of these two fragments may contain sequences which lie close to or at the borders of the T antigen cistron (30,31). Recent studies of Tegtmeyer, et al. (32) and Reed, et al. (33) strongly suggest that T antigen can negatively regulate its own synthesis by negatively regulating its own transcription. In that event, it is not unreasonable to speculate, by analogy with the effect of the λ C_I gene product on the transcription of the C_I gene (34,35), that the interaction of T with one or more sets of sequences within the SV40 early region might reflect a role for this protein affecting transcription.

Clearly, much additional information is needed before one can begin to gain solid appreciation for how T antigen functions in both lytic and transforming infection. The information presented here merely provides some preliminary suggestions whose sole virtue may prove to be that they are testable.

REFERENCES

(1) P. Tegtmeyer. J. Virol. 10 (1972) 591.

(2) J. Chou and R. Martin. J. Virol. 13 (1974) 1101.

(3) R. Martin and J. Chou. J. Virol. 15 (1975) 599.

(4) P. Tegtmeyer. J. Virol. 15 (1975) 613.

(5) J. Brugge and J. Butel. J. Virol. 15 (1975) 619.

(6) M. Osborn and K. Weber. J. Virol. 15 (1975) 636.

(7) S. Rozenblatt, R. Mulligan, M. Gorecki, B. Roberts, and A. Rich. Proc. Nat. Acad. Sci. (USA) 73 (1976) 2747.

(8) C. Prives, H. Aviv, B. Paterson, S. Rozenblatt, M. Revel, and E. Winocour. Proc. Nat. Acad. Sci. (USA) 71 (1974) 302.

(9) C.-J. Lai and D. Nathans. Virology 75 (1976) 335.

(10) J. Pope and W. Rowe. J. Exptl. Med. 120 (1964) 121.

(11) P. Black, W. Rowe, H. Turner, and R. Huebner. Proc. Nat. Acad. Sci. (USA) 50 (1963) 1148.

(12) F. Rapp, T. Kitahara, J. Butel, and J. Melnick. Proc. Nat. Acad. Sci. (USA) 52 (1964) 1138.

(13) M. Osborn and K. Weber. Cold Spring Harbor Symp. Quant. Biol. 39 (1974) 267.

(14) T. Kuchino and N. Yamaguchi. J. Virol. 15 (1975) 1302.

(15) D. Tenen, P. Baygell, and D. M. Livingston. Proc. Nat. Acad. Sci. (USA) 72 (1975) 4351.

(16) J. Alwine, S. Reed, J. Ferguson, and G. Stark. Cell 6 (1975) 529.

(17) D. Paulin and F. Cuzin. J. Virol. 15 (1975) 393.

(18) C. Prives, H. Aviv, E. Gilboa, E. Winocour, and M.Revel, in: In Vitro Transcription and Translation of Viral Genomes, Eds. A. L. Haenni and G. Beaud, INSERM pp. 305-312 (1975).

(19) R. Carroll and A. Smith. Proc. Nat. Acad. Sci. (USA) 73 (1976) 2254.

(20) R. Carroll, L. Hager, and R. Dulbecco. Proc. Nat. Acad. Sci. (USA) 71 (1974) 3754.

(21) D. Jessel, T. Landau, J. Hudson, T. Lalor, D. Tenen, and D. M. Livingston. Cell 8 (1976) 535.

(22) S. Reed, J. Ferguson, R. Davis, and G. Stark. Proc. Nat. Acad. Sci. (USA) 72 (1975) 1605.

(23) I. Henderson and D. M. Livingston. Cell 3 (1974) 65.

(24) D. Jessel, J. Hudson, T. Landau, D. Tenen, and D. M. Livingston. Proc. Nat. Acad. Sci. (USA) 72 (1975) 1960.

(25) R. Kelly, N. Cozzarelli, M. Deutsch, I. Lehman, and A. Kornberg. J. Biol. Chem. 245 (1970) 39.

(26) K. Danna, G. Sack, Jr., and D. Nathans. J. Mol. Biol. 78 (1973) 363.

(27) K. Danna and D. Nathans. Proc. Nat. Acad. Sci. (USA) 69 (1972) 3097.

(28) G. Fareed, C. Garon, and N. Salzman. J. Virol. 10 (1972) 481.

(29) J. Chou, J. Avila, and R. Martin. J. Virol. 14 (1974) 116.

(30) G. Khoury, P. Howley, D. Nathans, and M. Martin. J. Virol. 15 (1975) 433.

(31) R. Dhar, K. Subramanian, S. Zain, J. Pan, and S. Weissman. Cold Spring Harbor Symp. Quant. Biol. 39 (1974) 153.

(32) P. Tegtmeyer, M. Schwartz, J. Collins, and K. Rundell. J. Virol. 16 (1975) 168.

(33) S. Reed, G. Stark, and J. Alwine. Proc. Nat. Acad. Sci. (USA) 73 (1976) 3083.

(34) O. Kourilsky, M. Bourginon, M. Bouquet, and F. Gros. Cold Spring Harbor Symp. Quant. Biol. 35 (1970) 305.

(35) B. Meyer, D. Kleid, and M. Ptashne. Proc. Nat. Acad. Sci. (USA) 72 (1975) 4785.

(36) W. Hunter and F. Greenwood. Nature 194 (1962) 495.

(37) U. Laemmli. Nature 227 (1970) 680.

(38) A. Riggs, H. Suzuki, and S. Bourgeois. J. Mol. Biol. 48 (1970) 67.

ACKNOWLEDGEMENTS

This work was supported by grant CA 15751 from the National Cancer Institute. We are grateful to Ms. Ann E. Kennedy for her assistance in preparing this manuscript.

Discussion

A. Bollon, University of Texas: Is there any pattern in the GC content of these different fragments related to the binding?

D. Livingston, Harvard Medical School: We really do not know. Moreover, the fragments are rather large; actually constituting almost 50% of the genome. Therefore, until experiments with much smaller fragments are done, it will be impossible to say anything.

C. Weissmann, Universitet Zurich: Have you tried to do these binding experiments with chromatin complexes of the SV40? It seems as if this would be the more natural situation.

D. Livingston: Yes. Bob Martin and Bracielle DiLauro in Washington have carried out just such experiments and have observed T-chromatin complexes formed. Since Bob will speak this afternoon, I am sure that he will be happy to discuss his results with you in more detail at that time.

D. Hamer, Harvard Medical School: How well does T antigen bind to DNA other than SV40 DNA?

D. Livingston: T does bind to other DNAs. We have reported that it binds to a series of supercoiled phase DNAs. I think it will take more experimentation before we can define the relative affinities of SV40 vs. other DNAs for pure T antigen.

D. Hamer: How exactly do you calculate affinities from the sort of data that you have shown?

D. Livingston: We have used techniques of Ribbs in which DNA titrations are carried out with fixed, limiting amounts of T antigen. Dorothy Jessel and I carried out a series of salt and pH titrations and found that the conditions with which one can observe specificity of fragment binding were relatively particular. The NaCl concentration had to be approximately 0.066 or 0.07 molar, and the pH 6.8. We have never been able to observe any effect of manganese or magnesium on the specificity of binding.

K. Sakaguchi, Mitsubishi-Kasei Institute for Life Sciences, Tokyo: I would like to know whether the glycoprotein pattern in the membrane can be affected by the production of T antigen.

D. Livingston: I can only say this much: Bob Martin, Jeff Anderson, Peter Mora, Tony Chang and I have observed that T antigen has what we believe to be tumor specific transplantation antigen activity. Bob Martin will probably talk more about this later, but a highly purified T antigen fraction which is represented by only one major band in a gel also has TSTA activity. Now this and the finding by Gernot Walter and Bill Deppert that anti-T reactive polypeptides can be chased into plasma membrane fractions after synthesis in adeno-SV40 hybrid virus infected cells suggests that T antigen or some precursor or derivative may exist, at least in part, in the plasma membrane.

R. Kavenoff, University of California, San Diego: Is magnesium required for binding?

D. Livingston: No.

STUDIES OF ADENOVIRUS AND SV40 GENES REQUIRED FOR IN VITRO TRANSFORMATION

A.J.van der Eb, J.H.Lupker, J.Maat, P.Abrahams, H.Jochemsen, A.Houweling, W.Fiers[1)], C.Mulder[2)], H.van Ormondt and A.de Waard.

Sylvius Laboratories, University of Leiden, Leiden, the Netherlands.

1) Laboratory for Molecular Biology, University of Gent, Gent, Belgium.

2) Department of Pharmacology and Microbiology, University of Massachusetts, Medical School, Worcester, Massachu - setts 01605.

INTRODUCTION

The adenoviruses contain linear double-stranded DNA molecules with molecular weights ranging from 20-24 x 10^6 daltons. We have previously shown that the DNA of adenovirus type 5 is infectious and is able to induce in vitro transformation of rat cells (1,2). Further studies have shown that the transforming activity of adenovirus 5 (Ad5) DNA was unaffected when the DNA was degraded to segments as small as 1.5x10^6 daltons and that the activity was localized close to the left end of the molecule, starting at about 1% from this end and extending to about 6% (3).

As a further extension of these studies, attempts were made (a), to isolate the transforming genes of adenoviruses as unique DNA fragments, (b) to identify the proteins encoded by the transforming segments, and (c) to establish which of these proteins are required for transformation. In addition, we have started an analysis of the nucleotide sequence within the transforming region of Ad5 DNA, in order to obtain information on the structural organization of the transforming gene(s). The sequence of about 50 nucleotides in an area located at 4% from the left end of the genome will be presented. Furthermore, some preliminary results will be reported on the identification of transforming fragments of the DNA of oncogenic adenoviruses and of SV40,as well as on the possibility of testing the oncogenic potential of adenovirus DNA fragments directly in animals.

RESULTS AND DISCUSSION

Transforming fragments of adenovirus 2 and 5 DNA

As an extension of our studies on the localization of the transforming genes of human adenoviruses, DNA's of Ad2 and Ad5 were cleaved with several restriction endonucleases (endo's R) and the separated fragments were then tested for transforming activity,using the previously developed calcium technique(1,2)

Fig.1 shows the cleavage patterns of Ad2 and Ad5 DNA with

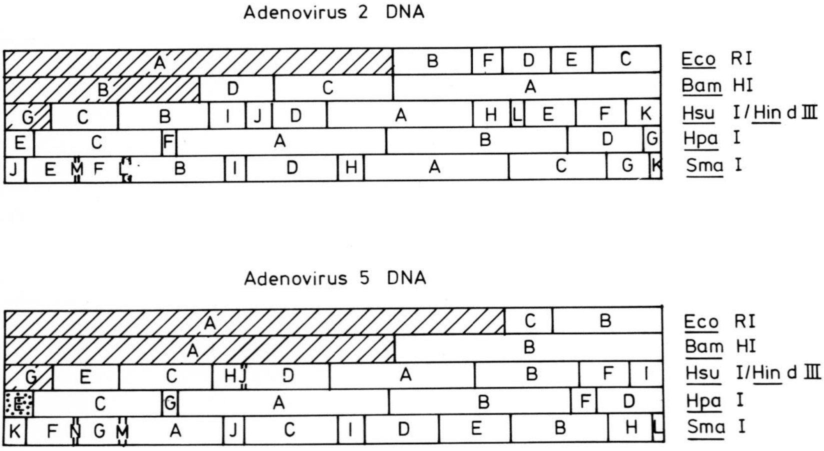

Fig.1. Maps of Ad2 and Ad5 DNA showing the cleavage sites of a number of restriction endonucleases. The hatched fragments were found to contain transforming activity. A low level of activity was found for Ad5 Hpa I E. The Eco RI and Hpa I maps are taken from (12), the Hin d III map of Ad2 DNA from R. Roberts (pers.comm.), the Bam H I and Sma I maps and the Ad5 Hsu I map from C. Mulder et al (pers. comm.) and (13)

the restriction endonucleases used in this work, and also summarizes our results obtained in the transformation assays. (Van der Eb et al, in preparation). It can be seen that fragments with transforming activity (hatched segments) were obtained with endo's R Eco R I, Bam H I and Hsu I and that these fragments all represent the leftterminal fragments of both viral DNA's. The smallest fragment with transforming activity is the Hsu I G fragment of both Ad2 and Ad5 DNA. This fragment represents about 7.3% of the viral genomes and is located at the extreme left end of the DNA's, in agreement with results obtained previously by us for Ad5 DNA (3). In early experiments no transforming activity was obtained with fragments produced with endo R. Hpa I and Sma I, indicating that these enzymes cleave into an area essential for transformation. Recent experiments have shown, however, that a fragment of Ad5 DNA, produced by cleavage with endo R. Hpa I did contain a low level of transforming activity, as will be discussed below.

Characterization of rat cells transformed by specific fragments of Ad2 and Ad5 DNA.

Several colonies of rat cells transformed by endo R. fragments of Ad2 and Ad5 DNA were established as cell lines, and were studied with respect to their ability to grow in semi-solid medium, as well as for the presence of adenovirus-specific T antigen.

All cell lines transformed by specific DNA fragments were found to grow to a very low level in medium containing 0.33% agarose. The cloning efficiency varied between 0.04% and 0.85% (occasionally lower than 0.04%), and there was no correlation between the size of the DNA fragment used to transform the cells and the ability to grow in the semi-solid medium (Van der Eb et al, in preparation).

The cell lines transformed by Ad2 and Ad5 DNA fragments were tested also for the presence of adenovirus subgroup C specific T antigen, using the indirect immunofluorescence technique with serum from Ad5-tumor bearing hamsters (4).The cell lines transformed by the Eco R I A and Bam H I fragments were found to contain a fluorescence pattern basically similar to the pattern observed in the cells transformed by intact virus: one or more fluorescent spots or granules in the nucleus and often in the area surrounding the nuclear membrane (Fig. 2a). However, a different and atypical pattern was observed in the cells transformed by the 7% Hsu I G fragment of Ad5 DNA (cells transformed by Ad2 Hsu I G have not been studied). All G-fragment transformed lines were found to react with anti T-serum, but the distribution of the fluorescence was abnormal in that it was predominantly concentrated in the cytoplasm as a granular or diffuse staining (Fig.2 b and c). However, when three subclones of Hsu I G-transformed clone VI, derived from colonies growing in 0,33% aga-

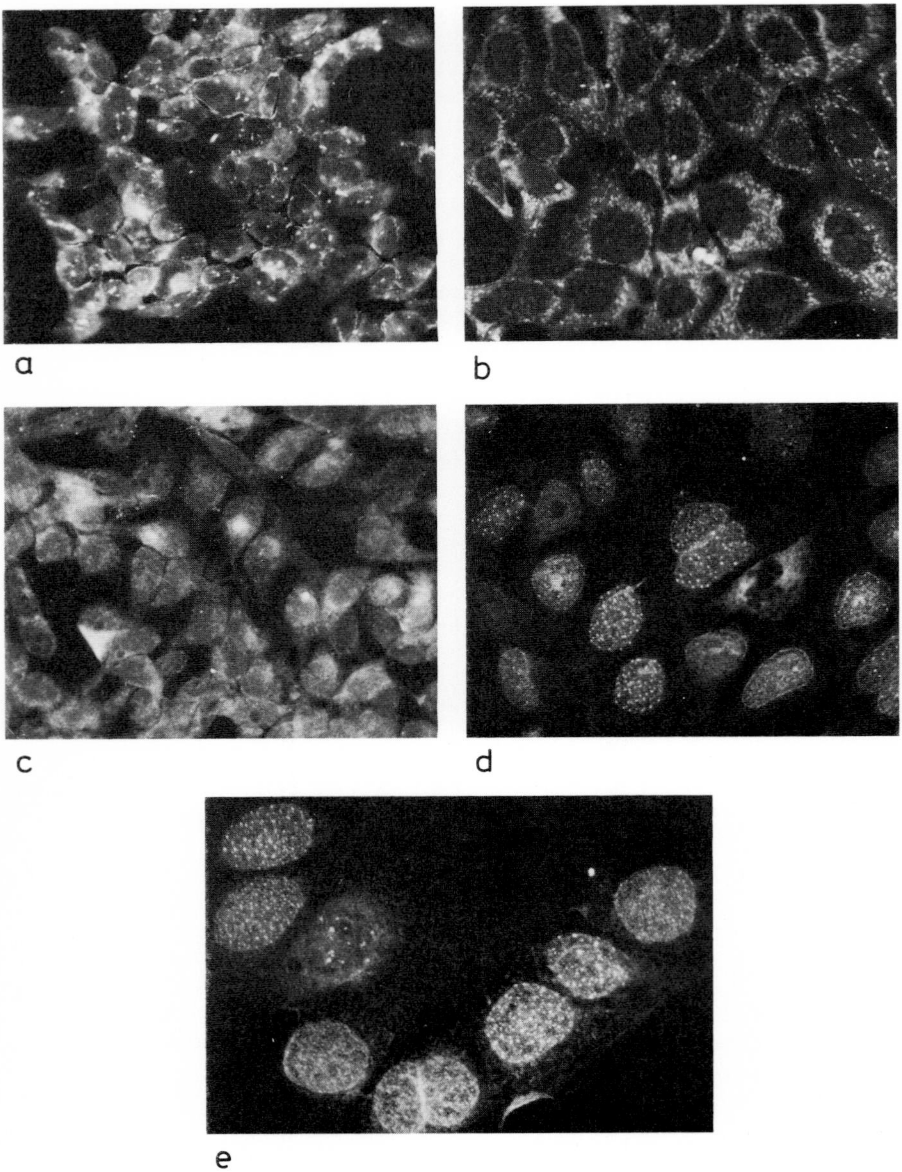

Fig.2. T antigen staining patterns of rat kidney cells transformed by adenovirus DNA fragments. The cells were transformed by (a) Ad2 Eco RI A; (b) Ad5 Hsu I G, clone I; (c) Ad5 Hsu I G, clone VI; (d) and (e) Ad5 Hsu I G clone VI, subclones 2 and 1 respectively, derived from colonies growing in 0.33% agarose.

rose medium, were analyzed for T antigen, it was found that the distribution of the fluorescence again differed from the original clone VI (Fig.2 d and e). All three subclones contained numerous fluorescent flecks in 50-70% of the nuclei, a pattern resembling the typical adenovirus T antigen distribution. We concluded on the basis of these results that the Hsu I G transformed cells indeed contain adenovirus-specific T antigen, but that its distribution is abnormal (Van der Eb et al, in preparation).

Viral DNA sequences in the Hsu I G transformed cells.

In order to find out whether the cell lines transformed by the 7% Hsu I G fragment of Ad5 DNA contain viral genetic information five clones of Hsu I G-transformed rat cells were analyzed for the presence of viral DNA sequences in their cellular DNA. In these analyses, the renaturation kinetics of ^{32}P labeled endo R· fragments of Ad5 DNA was measured in the presence of large amounts of DNA extracted either from the transformed cells, or from untransformed cells, essentially as described by Gallimore et al(5).

DNA from Hsu I G transformed cells was analyzed with ^{32}P labeled fragments of Ad5 DNA obtained by cleavage with endo R·Hsu I. Fig.3 shows that this enzyme cleaves Ad5 DNA into 10 fragments. To reduce the number of assays, the adjacent fragments D and H, as well as F and I were tested as mixtures.

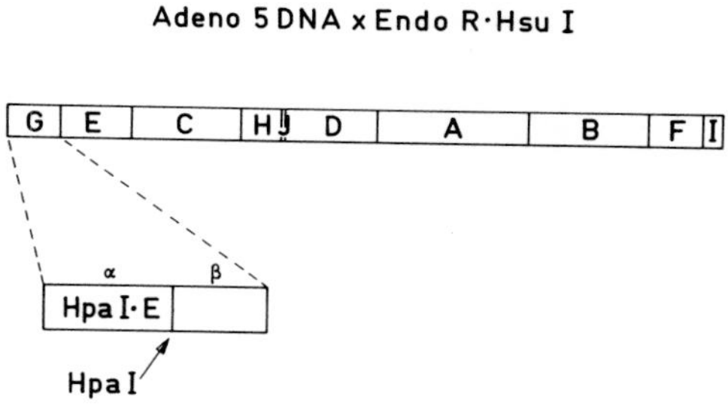

Fig.3. Map of Ad5 DNA showing the cleavage sites of endo R. Hsu I. Fragment Hsu I G is cleaved by endo R. Hpa I into the 2 fragments designated α and β .

The results obtained with DNA of one of the G-fragment transformed clones are illustrated in Fig.4. It can be seen

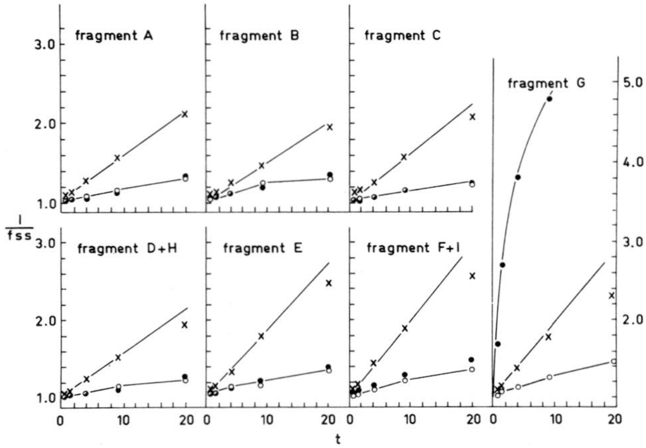

Fig.4. Renaturation kinetics of ^{32}P labeled Hsu I fragments A, B, C, D+H, E, F + I and G of Ad5 DNA, in the presence of DNA from Ad5 Hsu I G-transformed clone VI (●), DNA from calf thymus (○), or DNA from calf thymus containing unlabeled Ad5 DNA (X). The reaction mixtures contained 1000 µg/ml of cellular DNA, and 4000 cpm/ml of one of the ^{32}P labeled DNA fragments. The reaction mixtures of the reconstruction experiment contained 1000 µg/ml calf thymus DNA, 4000 cpm/ml of one of the ^{32}P labeled DNA fragments, and 1.2×10^{-2} µg unlabeled Ad5 DNA /ml which corresponds to 2 copies per diploid cell. The DNA's were degraded to segments of about 300 nucleotides by boiling in alkali and renaturation was carried out in 0.8 M sodium phosphate buffer pH 6.8 at 68°; the fraction of single-stranded DNA was determined by hydroxylapatite chromatography (see ref.5).

that the ^{32}P labeled fragments reannealed slowly in the presence of calf thymus DNA (used here as untransformed cell DNA) but that the rate of renaturation of all fragments was considerably increased when a small amount of unlabeled Ad5 DNA (about 2 copies per cell) was added to the calf thymus DNA (reconstruction experiment). Furthermore, it can be seen that the ^{32}P fragments A, B, C, D+H, E and F+I renatured at the same rate in the presence of either transformed cell DNA or calf thymus DNA, but that the rate of reannealing of fragment G was strongly increased in the presence of transformed cell DNA. Hence, the transformed cells contained sequences homologous to Hsu I fragment G and to no other fragment. From the kinetics of renaturation of ^{32}P labeled fragment G, it was calculated that the transformed cells contained approximately 35 copies of fragment G per amount of diploid cell DNA, and that the sequences of fragment G were present for about 85%. Basically similar results were obtained with the DNA's isolated from Hsu I G transformed clones II-V (Van der Eb et al, in preparation).

To obtain a more precise estimate as to which part of the Hsu I G sequence was present and which part was absent, the DNA's from the Hsu I G transformed clones II,III,V and VI were also analyzed with the two ^{32}P labeled fragments, designated α and β, which arise when fragment Hsu I G is cleaved with endo R. Hpa I (see Fig.3). The results demonstrated that ^{32}P labeled fragments α and β followed renaturation kinetics in the presence of the transformed cell DNA's similar to that of ^{32}P labeled Hsu I G, indicating that both α and β sequences were present in the cells in several copies, and to the extent of 73-85% of their length. A summary of the results of the renaturation kinetic analyses are presented in table 1.

Assuming that the missing 15-25% of fragment G are deleted from the ends of this fragment, then the Ad5 sequences present in clones II,III,V and VI would be distributed as shown in fig.5.

It was concluded from these results that the Ad5 Hsu I G transformed cells do contain sequences homologous to 75-85% of this fragment, indicating that a segment as small as 5.5-6% of the viral genome ($1.3-1.4 \times 10^6$ daltons) contains information for the maintenance of transformation.

Identification of the proteins encoded by the transforming segment of Ad5 DNA.

The studies described above have demonstrated that the left-most 7% of Ad5 DNA contains the information for the induction and the phenotypic expression of transformation. The abnormal distribution of the T antigen(s) in the Hsu I G-transformed cells suggested, however, that a viral encoded protein, which normally is involved in transformation, was defective or absent in these cells. In order to identify the proteins involved in transformation, virus-specific RNA homologous to the transforming DNA segment was isolated and translated in a wheat germ protein synthesizing system. The RNA was isolated from the cytoplasm of Ad5-infected KB cells in the

TABLE 1

Characterization of five rat cell lines transformed by Ad5 Hsu I fragment G, with respect to the number of copies, and the fraction of fragments Hsu I G, α or β, present per quantity of diploid cell DNA. The results are obtained from reassociation kinetic analyses as described in the text.

Clone number	^{32}P labeled fragment used for analysis	Number of copies per diploid cell	% of fragment present in cells
II and V	fragment G	48	80
	α	48.5	85
	β	32	74
III	fragment G	10	76
	α	3.5	78
	β	5.5	73
IV	fragment G	22	75
VI	fragment G	34	84
	α	36	88
	β	24	78

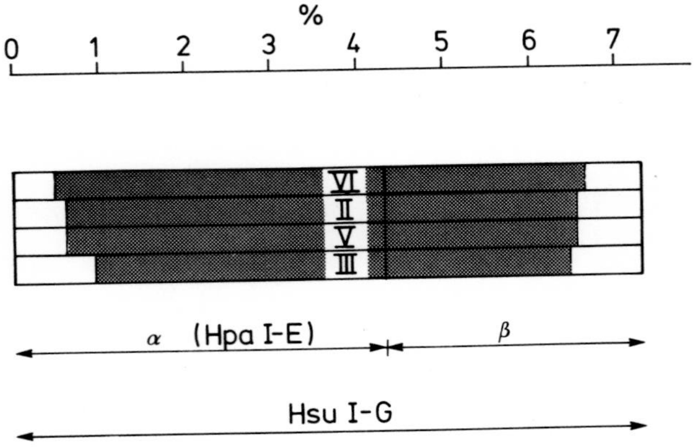

Fig.5. Sequences of Ad5 fragment Hsu I G and fragments α and β, estimated to be present in Hsu I G transformed clones II, III, V and VI (shaded areas), assuming that the missing sequences are deleted from the ends of fragment G (see also table 1).

early phase of the infection, or from Hsu I G transformed cells. The virus-specific RNA molecules were selected from the total cytoplasmic RNA by hybridization to viral DNA or DNA fragments, bound to nitrocellulose filters.

Fig.6 summarizes the results obtained with RNA's selected by hybridization to total Ad5 DNA, and to a number of DNA

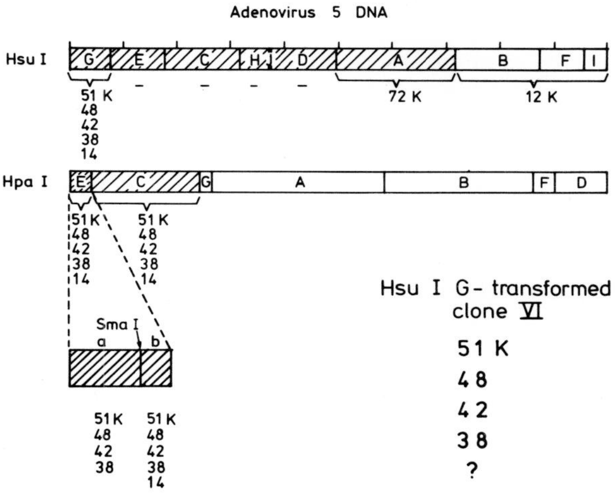

Fig.6. Proteins synthesized in a wheat germ cell-free system, using mRNA selected by hybridization to Ad5 DNA or to Ad5 DNA fragments (hatched fragments) obtained by cleavage with endo's R. Hsu I and Hpa I. RNA was also selected with the two fragments, designated a and b, obtained by cleavage of Hpa I E with endo R. Sma I. The proteins synthesized in response to virus-specific RNA isolated from Hsu I G-transformed clone VI are also shown.

fragments. RNA selected by hybridization to total Ad5 DNA was translated into at least 7 polypeptides. Five of these polypeptides (51K, 48K, 42K, 38K and 14K) were encoded by the 7% Hsu I G fragment while the remaining two originated from the right part of the genome. Basically similar results were reported by Lewis et al (6) who obtained 44-50K and 15K polypeptides with RNA homologous to the left terminal DNA sequences of Ad2 DNA.

To obtain more information on the localization of the genes coding for the Hsu I G-specified proteins translation experiments were carried out also with RNA selected with Ad5 Hpa I fragments E and C (which both overlap with Hsu I G) and with the two fragments, designated a and b, obtained by cleavage of Hpa I E with endo R. Sma I. Fig.6 shows that the RNA's selected with Hpa I E and C and with fragment b, were translated into the same 5 proteins as were obtained with Hsu I G, but that fragment a-selected RNA probably was translated into the 4 largest proteins only.

The results suggested the following: (1) Some of the 5 polypeptides synthesized on Hsu I G- selected RNA must be encoded by common DNA sequences and hence may be partially identical, since the maximum coding capacity of Hsu I G corresponds to only 100.000 daltons of protein. (2) The 14K polypeptide is synthesized on an RNA species which maps in an area at the right hand side of the Sma I cleavage site (3)The RNA coding for the 14K polypeptide, and the (one or more) RNA species coding for the 4 larger polypeptides, should be overlapping in the area of the Hpa I cleavage site.Preliminary experiments indicated that Ad5 specific RNA isolated from the Hsu I G-transformed clone VI was translated into the 4 larger polypeptides of 51,48,42 and 38K, while the 14K polypeptide could not be identified (Fig.6). This result suggests that the abnormal T antigen pattern of the G fragment-transformed cells might be related to the absence or to defectiveness of the 14K polypeptide. Further studies are in progress on the relationship between the various proteins encoded by the Hsu I G fragment. In this connection, it is of interest that in a protein synthesizing system derived from ascites cells, Ad5 specific early RNA is translated into the same 7 polypeptides as in the wheat germ system. This indicates that the synthesis of the 5 different polypeptides with Hsu I G selected RNA is not an artifact of the wheat germ system.

Isolation of a 4% fragment of Ad5 DNA with transforming activity

As stated above, transforming activity could not be detected in early experiments with the fragments of Ad2 and Ad5 DNA produced by cleavage with endo R. Hpa I. The enzyme preparation used in these experiments, however, was shown to be contaminated with some exonuclease activity, and this might well have destroyed the transforming activity. Recently, we isolated a Hpa I enzyme preparation free of detectable exonuclease activity. When the fragments produced by this enzyme were tested, it was now found that a low level of transforming activity was associated with Hpa I fragment E, which represents the left terminal 4.3% of Ad5 DNA. The specific transforming activity of Hpa I E was consistently lower, however, (at least 5-fold) than that of the previously identified fragments. A possible explanation for this low activity may be that one of the "ends" of a viral gene essential for transformation is located close to the Hpa I cleavage site.Many E fragments which would normally be able to cause transformation, may then be inactivated by cellular exonucleases or during the integration process (which often seems to result in a deletion of terminal DNA sequences). To investigate whether the Hpa I E fragment-transformed cells contained DNA sequences homologous to Ad5 Hpa I E, two transformed cell lines were analyzed by reassociation kinetics for the presence of Ad5 DNA sequences. The results demonstrated that both lines clearly contained such sequences(one line contained about 2 copies and the second line 15-20 copies of the Hpa I E fragment per cell). This supports our conclusion that the Hpa I E fragment does contain transforming activity. Experiments are in progress to analyze a number of other Hpa I E fragment-transformed lines for the presence of viral DNA sequences, and to study the growth properties and the T antigen distribution of the cells.

Nucleotide sequence analysis in the transforming region of Ad5 DNA

The low level of transforming activity observed with the Hpa I E fragment could be due to the localization of the Hpa I cleavage site (at 4%) very closely to a sequence coding for the transforming protein. To obtain more precise information on the structural organization of this region of the genome, we have started an analysis of the nucleotide sequence in the right hand part of the fragment Hpa I E.

A DNA segment suitable for sequence analysis was obtained by degrading the Hpa I E fragment with the restriction endonucleases Hpa II and Hae III. The resulting fragments were ordered on a physical map as shown in Fig.7 (J.Maat et al, unpublished results)

Sofar we have determined the primary structure of the segment between the Alu I and the Hpa I cleavage sites. The Hpa I E fragment was labeled by treatment with exonuclease III for various lengths of time, followed by a repair reaction with α ^{32}P-labeled dNTP's and DNA polymerase I. Upon degradation with endo R. Alu I and Hae III, the nucleotide sequences were determined by a two-dimensional homochromatography of partial snake venom exonuclease digests (7).

Corroborative evidence was obtained in a number of ways, among which pyrimidine tract analysis and 5'- labeling with T4-induced polynucleotide kinase. We also applied Sanger and Coulson's "plus-minus" technique (8). The heterogeneous 3'-termini required for this method were obtained by exonuclease III treatment which, upon one single incubation with DNA polymerase and labeled dNTP's, gives the appropriate substrate for the "plus-minus technique ". The Hpa I E fragment was then

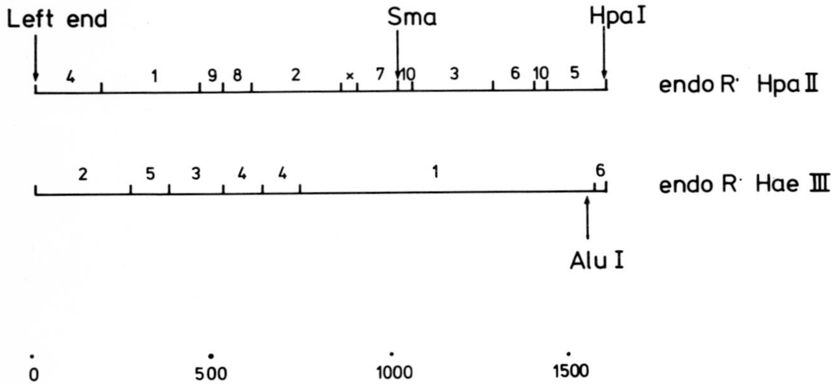

Fig.7. Cleavage maps of fragment Hpa I E of Ad5 DNA by endo's R. Hpa II and Hae III.

cleaved with endo R. Sma I, in order to remove the left-hand part of the fragment. The right hand part was subjected to the "plus-minus" incubations and subsequently treated with endo R. Hae III. An advantage of this approach is that double-stranded DNA will become accessible to sequencing according to the "plus-minus" technique, without prior strand separation and isolation of single strands.

The nucleotide sequence deduced is represented in Fig.8. The gap in this sequence is a site where an ambiguity is still to be resolved. Philipson et al.(9) and Sharp et al (10) have reported that transcription proceeds to the right in the corresponding region of the genome of the closely related Ad2; consequently, the sequence given in Fig.8 must be identical to the transcription product. In our sequence one T-A-A stop-codon occurs in each of the possible three reading frames. Therefore, no translation is possible beyond the right most T-A-A- triplet. In view of the low efficiency of transformation by the Hpa I E fragment and the occurrence of these three stop-codons in the vicinity of the Hpa I cleavage site, one of these T-A-A triplets may well be the termination signal of the transforming protein.

Further sequence studies in this area may reveal whether any of these reading frames corresponds to a coding sequence for the transforming protein.

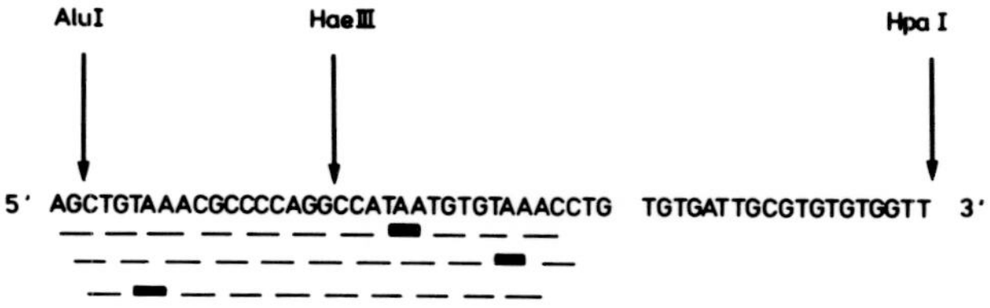

Fig.8. Nucleotide sequence at the right side terminus of Hpa I-E. The fragment represented by this sequence was obtained by cleavage of Hpa I-E by endo R. Alu I.

Transforming activity of Ad12, 3 and 7 DNA, and oncogenic activity of Ad12 DNA

Experiments similar to those described for the nononcogenic adenoviruses 2 and 5 have also been started with oncogenic Ad12, and recently with Ad3 and Ad7. Intact DNA of these adenoviruses contains transforming activity although the efficiency of transformation is at least 5 fold lower than that of Ad2 and Ad5. DNA's of the oncogenic Ad3, 7 and 12 were cleaved with several restriction endonucleases and each of the fragments were tested for transforming activity. Preliminary results indicate that a terminal fragment of Ad3 and Ad7 DNA, obtained by cleavage with endo R. Sma I contains transforming activity. In addition, transformation was obtained with the Eco RI C- fragment of Ad12 DNA (16% of the viral genome). We are now attempting to isolate smaller transforming fragments of Ad12 DNA.

An interesting recent observation is that DNA of Ad12 can also induce tumors directly in animals. Injection of newborn hamsters with 4 to 5 μg Ad12 DNA or with an equivalent amount of Eco RI fragment C resulted in the induction of a small number of tumors. Analysis by reassociation kinetics demonstrated that a tumor induced by intact DNA contained most of the Ad12 DNA sequences, and that an Eco RI C-induced tumor contained only sequences homologous to fragment C in several copies. This result shows that it is also possible to test the oncogenic potential of DNA fragments in animals.

Transforming fragments of SV40 DNA

Abrahams et al. recently reported the isolation of 2 restriction endonuclease fragments of SV40 DNA with transforming activity. Both DNA fragments contained the entire early region as well as the origin of DNA replication(11).

The fragments are the 74% Eco RI/Hap II A fragment and the 59% Bam HI /Hap II A fragment (Fig.9). Analysis of the viral DNA sequences present in the cellular DNA by reassociation kinetics with ^{32}P labeled SV40 DNA fragments have shown that the transformed cells only contained sequences homologous to the DNA fragments used to transform the cells. Fragments obtained by cleavage with endo R. Hpa I, which introduces one break in the early region, did not contain transforming activity. Recent observations suggested that a fragment prepared with endo R. Eco RI and Bgl I (Bgl I cleaves in the area of

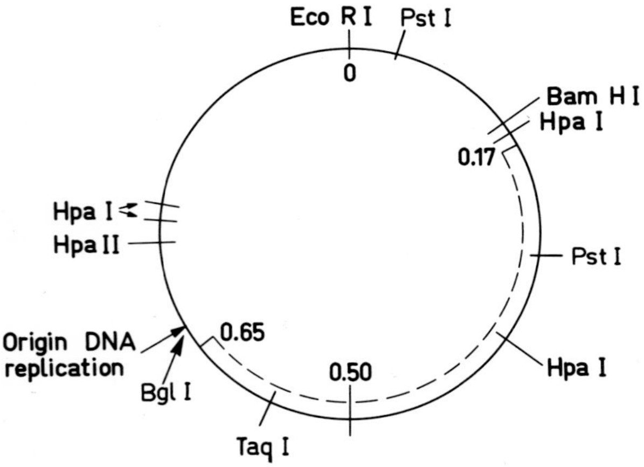

Fig.9. Map of SV40 DNA showing the cleavage sites of endo's R. Eco RI, Bam HI(Mulder and Greene, unpublished result), Pst I (14), Hpa I and Hpa II (15), Taq I (16) and Bgl I (16). The part of the genome which is expressed in the early phase of the infection is indicated with a broken line.

the DNA-replication origin) has transforming activity.

Two additional restriction endonucleases (Pst I and Taq I) have been used recently to investigate the possibility of isolating transforming fragments which contain only a part of the early region. Preliminary results indicated that fragments prepared with endo R. Pst I (this enzyme cleaves at 4% and at 27%) did not contain transforming activity, but that transformation could be obtained with the Eco RI/Taq I A fragment (Taq I cleaves SV40 DNA at 57%). This suggests that the segment of the early region between 57% and the origin of DNA replication is not strictly required for transformation. Experiments are in progress to analyse these transformed cells for the presence of viral DNA sequences and to study their T antigen pattern.

Acknowledgements

We thank K. Sol for providing us with a number of restriction endonucleases. The expert technical assistance of Arja Davis, Mieke Wille and Jacqueline Hertoghs is gratefully acknowledged.

This work was supported in part by the Netherlands Organization for the Advancement of Pure Research (ZWO), through funds from the Foundation for Medical Research (FUNGO) and the Foundation for Chemical Research(SON).

W.F. is supported by the Kankerfonds van de Algemene Spaar- en Lijfrentekas (Belgium) and C.M. by Cancer Society Grant V.C. 197.

REFERENCES

(1) F.L. Graham and A.J. van der Eb, Virology 52 (1973) 456.

(2) F.L. Graham and A.J. van der Eb, Virology 54 (1973) 536.

(3) F.L. Graham, A.J. van der Eb and H.L. Heijneker, Nature 251 (1974) 687.

(4) F.L. Graham, P.J. Abrahams, C. Mulder, H.L. Heijneker, S.O. Warnaar, F.A.J. de Vries, W. Fiers and A.J. van der Eb, Cold Spring Harbor Symp. Quant. Biol. 39 (1974) 637.

(5) P.H. Gallimore, P.A. Sharp and J. Sambrook, J.Mol. Biol. 89 (1974) 49.

(6) J.B. Lewis, J.F. Atkins, P.R. Baum, R. Solem, R.F. Gesteland and C.W. Anderson, Cell 7 (1975) 141.

(7) F. Sanger, J.E. Donelson, A.R. Coulson, H. Kössel and D. Fischer, Proc. Natl. Acad. Sci. U.S.A. 70 (1973) 1209.

(8) F. Sanger and A.R. Coulson, J. Mol. Biol. 94 (1975) 441.

(9) L. Philipson, U. Pettersson, U. Lindberg, C. Tibbetts, B. Vennström, T. Persson, Cold Spring Harbor Symp. Quant. Biol. 39 (1974) 447.

(10) P.A. Sharp, P.H. Gallimore and J. S. Flint, Cold Spring Harbor Symp. Quant. Biol. 39,(1974) 457.

(11) P.J. Abrahams, C. Mulder, A. van de Voorde, S.O. Warnaar and A.J. van der Eb, J. Virol. 16 (1975) 818.

(12) C. Mulder, J.R. Arrand, H. Delius, W. Keller, U. Pettersson, R.J. Roberts and P.A. Sharp, Cold Spring Harbor Symp. Quant. Biol. 39 (1974) 397.

(13) J.S. Sussenbach and H.G. Kuijk, Virology, in press.

(14) D.I. Smith, F.R. Blattner and J. Davies, Nucl. Acid. Res. 3 (1976) 343.

REFERENCES (continued)

(15) K.J. Danna, G.H. Sack and D. Nathans, J. Mol. Biol. 78 (1973) 363.

(16) B.S. Zain and R.J. Roberts, unpublished observations; quoted in Critical Rev. Bioch. 4 (1976) 123.

SV40-ADENOVIRUS 2 HYBRID VIRUSES

J. Sambrook
Cold Spring Harbor Laboratory
P.O. Box 100
Cold Spring Harbor, New York 11724

Abstract: A new set of hybrid viruses has been isolated whose closed circular genomes 5-6 KB in size, contain DNA sequences derived in part from adenovirus 2 and in part from SV40. The structure of arrangement of these genomes is complex but in the simplest case, analyses by restriction endonuclease digestion and hybridization indicate that the adenovirus 2 DNA is present as a continuous block, of maximum size 2.8 KB. Different hybrids contain sequences derived from different segments of the adenovirus 2 genome.

INTRODUCTION

The idea that segments of foreign DNA can be propagated in mammalian cells as part of the SV40 genome was first raised in 1972 by Lavi and Winocour (1). They showed that serial passage of the virus in permissive cells leads to the appearance of genomes that consist of both host and viral DNA sequences. In the simplest case, part of the SV40 genome is replaced by monkey DNA of either the repetitive or unique classes (2, 3, 4) to creat a defective genome that can multiply only in the presence of a helper virus. Subsequently Fareed and his colleagues (5) used <u>in vitro</u> ligation to join a selected piece of bacteriophage λ DNA to a defined segment of the SV40 genome. The resulting chimeras were introduced into monkey cells together with wild-type SV40 DNA, which provided all the gene products required to replicate and pack the hybrid DNA molecules into virus particles. In principle, it seems possible to use this system and variants of it to amplify any desired segment of DNA that, together with its vector, is small enough to be packed into an SV40 virion.

In this paper, I describe the isolation of hybrid viruses whose defective genomes consist partly of SV40 DNA and partly of the DNA of the unrelated virus, adenovirus 2. Unlike Fareed, I have not used <u>in vitro</u> ligation but have relied upon cellular recombination enzymes during the construction of these hybrids; by contrast to Lavi and Winocour however I have been able to select for hybrid molecules of the desired kind.

The isolation scheme is shown in Figure 1. Ad2$^+$D1 is a defective adenovirus 2-SV40 hybrid that can be propagated only in the presence of a helper adenovirus: under these conditions it grows well and forms a surprisingly high proportion (40%) of the virus yield (6). Its genome contains an insertion of SV40 DNA about 2KB in length, in place of the 3.5 KB segment of adenovirus 2 DNA which maps between 0.64 and 0.74 fractional genome lengths and includes the gene for the 72 K DNA-binding protein. The inserted SV40 sequences begin at position 71 on the conventional map of that virus' genome. Stretching away to the left are the regions that code successively for the origin of replication, the A gene, the terminator at position 17 and finally part of the gene for the major coat protein, VP1: the integrated SV40 sequences end at position 11 (6) (see Figure 2). During lytic infection by Ad2$^+$D1 the integrated SV40 DNA sequences are transcribed and translated into a protein which can be precipitated by SV40-specific anti-T serum and whose mobility through SDS-polyacrylamide gels is indistinguishable from that of authentic T antigen - the product of the early SV40 A gene (7). On the basis of this evidence, it seemed entirely possible that Ad2$^+$D1 might complement _ts_ mutants of the SV40 A gene. Because multiplication of SV40 is efficiently suppressed in cells infected by adenovirus 2 (8), the simplest and most direct test for complementation is coinfection of cells with DNA fragments, using the techniques developed by Mertz and Berg (9). Accordingly the DNA of Ad2$^+$D1 was cleaved with one of a variety of restriction endonucleases and introduced into monkey cells at non-permissive temperature together with intact DNA of SV40 _ts_ a 30 (10). Amongst the resulting progeny were viruses whose closed circular genomes contained adenovirus 2 DNA sequences.

EXPERIMENTAL

The origin and procedures for the growth of the CV-1 line of monkey kidney cells have been described (11). Plaque assays with SV40 virions or DNA were performed as described by Mertz and Berg (9). Ad2$^+$D1 (6), isolated from a stock of Ad2^{++}HEY (12), was propagated in CV-1 cells. The virus particles were purified and DNA extracted from them as described by Pettersson and Sambrook (13). SV40 _tsa_ 30, obtained from Dr. Peter Tegtmeyer (10) was grown at 32°C in CV-1 cells which had been infected at a multiplicity of 0.05 plaque-forming units per cell.

SV40 tsa 30 DNA was extracted by the method of Hirt (14) from CV-1 cells infected at a multiplicity of one plaque-forming unit per cell, after 72 hours incubation at 32.5°C.

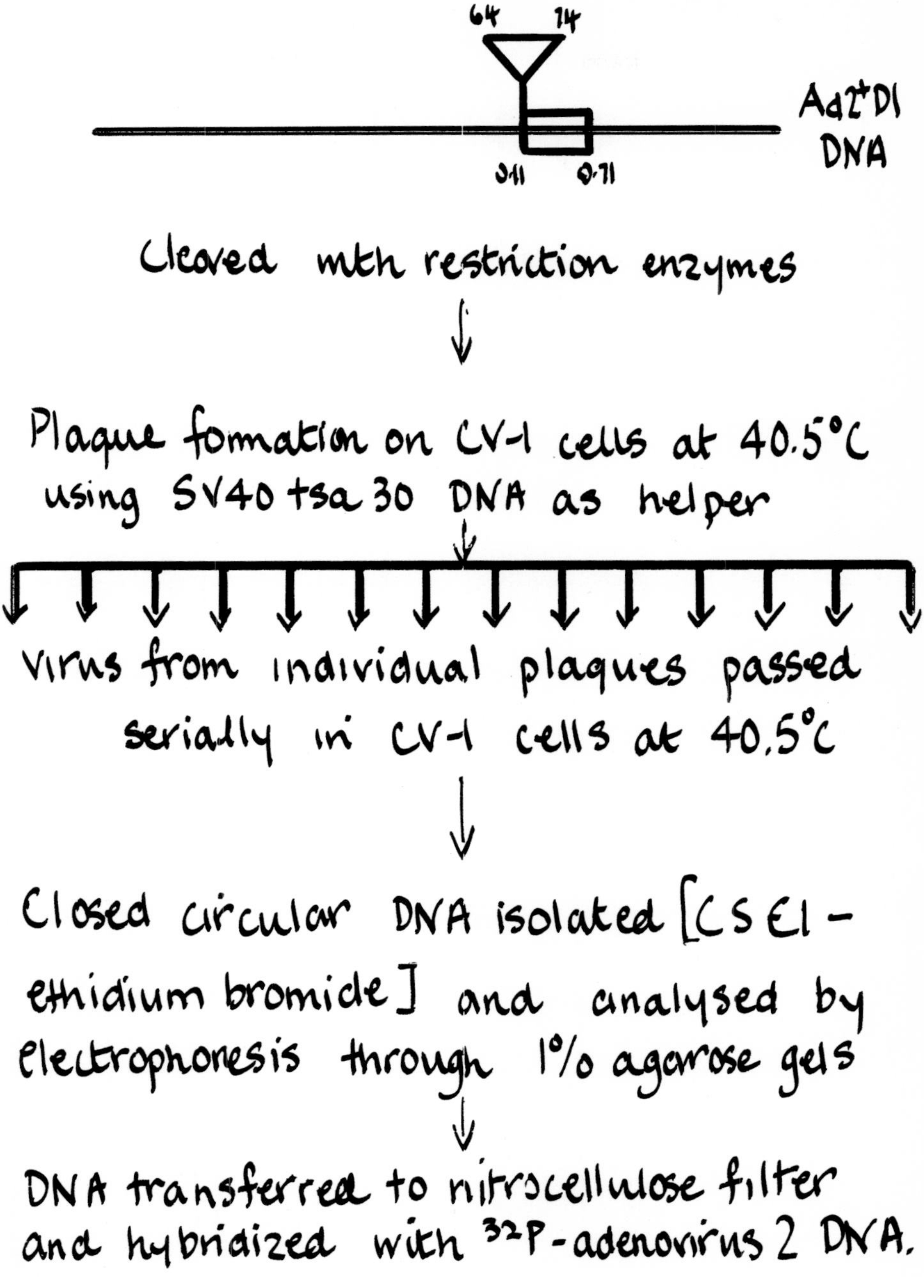
64
14
Ad2+D1
DNA
0.11
0.71
Cleaved with restriction enzymes
Plaque formation on CV-1 cells at 40.5°C
using SV40 tsa 30 DNA as helper
Virus from individual plaques passed
serially in CV-1 cells at 40.5°C
Closed circular DNA isolated [CsCl-
ethidium bromide] and analysed by
electrophoresis through 1% agarose gels
DNA transferred to nitrocellulose filter
and hybridized with 32P-adenovirus 2 DNA.

Figure 1. Isolation scheme for SV40-adenovirus 2 hybrids.

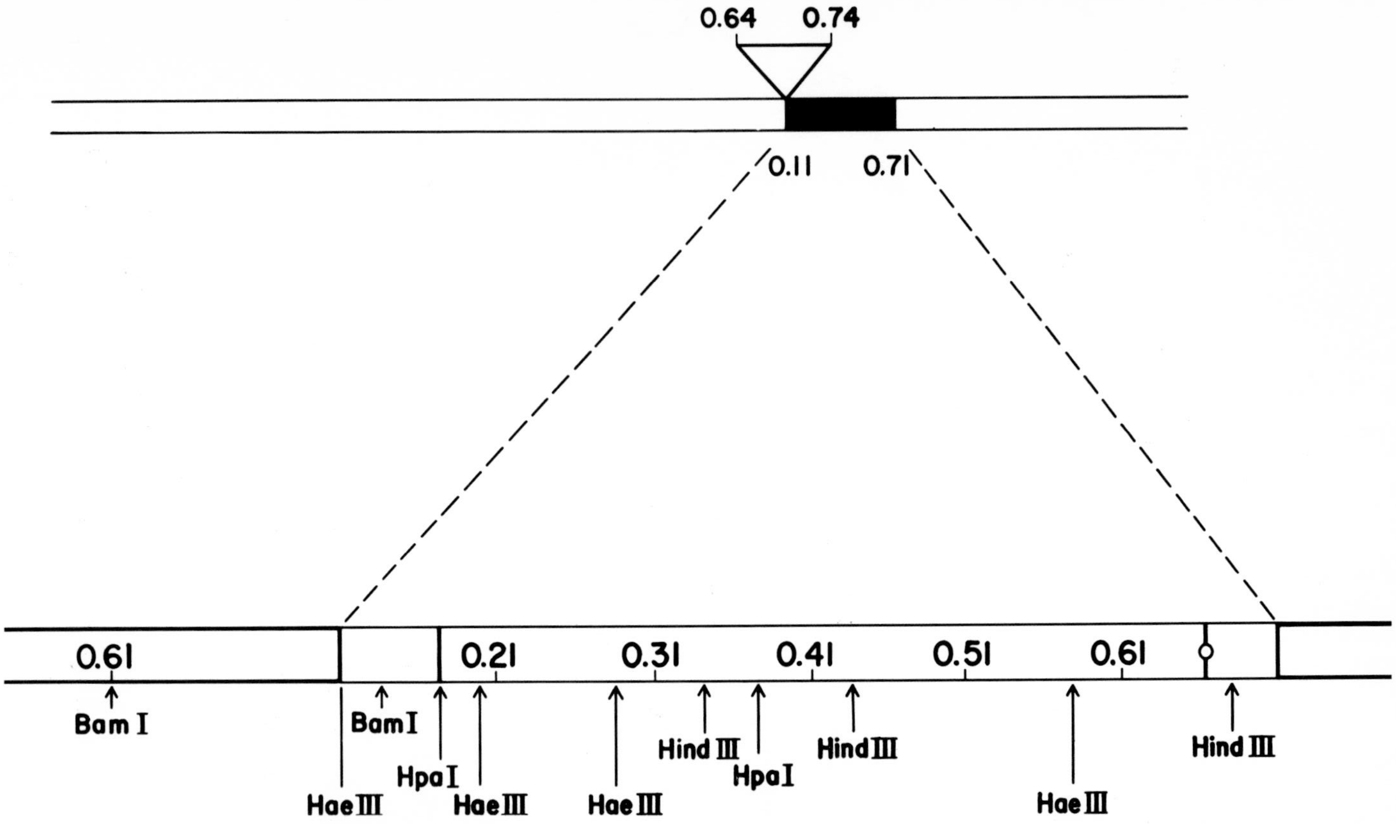

Figure 2. Structure of Ad2$^+$D1 DNA. The top diagram shows the location of the inserted SV40 DNA sequences and the site at which adenovirus 2 DNA sequences are deleted. The bottom diagram shows a map of the sites at which restriction endonucleases cleave the integrated SV40 DNA. Data of Lukanidin and Sambrook.

Covalently closed viral DNA was purified by centrifugation to equilibrium in a solution containing CsCl 1.59 g/cc) and ethidium bromide (200 μg/ml). After the band of component I DNA was collected the ethidium bromide was removed by extraction with isopropanol. Endonucleases Eco RI, Hpa I, Sma I, Hha I, Hind III and Bgl I were prepared and used as described previously (15).

Agarose gels (1%) were prepared in Tris-acetate buffer (0.04 M Tris pH 7.8, 0.005 M sodium acetate, 0.001 M EDTA). Samples were applied in 50 μl of Tris acetate buffer containing sucrose (20% weight/volume) and electrophoresis was carried out for 20 hours at 40V (16). DNA bands were stained with ethidium bromide and photographed as originally described by Sharp *et al*. (17).

For hybridization experiments, bands of DNA in agarose gels were denatured *in situ* and transferred to nitrocellulose filters according to Southern (18). Introduction of α-^{32}P nucleotides into DNA by "nick-translation" was developed by Berg and his colleagues as a method to generate hybridization probes (19): in this work the conditions established by Maniatis *et al* (20) were used. Hybridization and autoradiography was carried out exactly as described (21).

This work was carried out in laboratory designed, equipped and maintained to standards higher than those required by N.I.H. for the handling of adenovirus 2-SV40 hybrids. In addition all manipulation of infectious material was carried out in the deep hours of the night when few other people were around, in order to minimize the chances of exposure in the event of an accidental spill. No such calamity in fact occurred.

RESULTS

Complementation between fragments of Ad2$^+$D1 DNA and SV40 tsa 30

Ad2$^+$D1 DNA was cleaved with various restricting endonucleases and the resulting fragments were purified mixed with SV40 tsa 30 DNA and used to infect monolayers of CV-1 cells. Table 1 shows that the number of plaques obtained after 12 days incubation at 40.5°C varies with the dose of Ad2$^+$D1 fragments applied to the cells and with the particular restriction enzyme employed. No plaques were obtained when SV40 *tsa* 30 DNA was used alone or in conjunction with either intact Ad2$^+$D1 DNA or fragments of it derived by digestion with endonucleases Hpa I and Hind III - enzymes which cleave within the sequences of SV40 gene A that are integrated into the Ad2$^+$D1 genome. By contrast, coinfection of cells with SV40 *tsa* 30 DNA and with fragments obtained by digestion

Complementation between fragments of $Ad2^{+}D1$ DNA and SV40 tsa 30 DNA

Restriction endonuclease	Amount of fragments of $Ad2^{+}D1$ DNA (microgram)	Amount of SV40 tsa 30 DNA (microgram)		
		0	0.1	0.5
0	0	0	0	0
	1.0	0	0	0
	5.0	0	0	0
	10.0	0	1	0
Sma I	0	0	1	0
	0.1	0	1×10^{1}	4×10^{1}
	0.5	0	8×10^{1}	2×10^{1}
	5.0	0	8×10^{2}	4×10^{2}
Bgl 1	0	0	0	0
	0.1	0	6×10^{1}	3×10^{1}
	0.5	0	6×10^{1}	11×10^{1}
	5.0	0	12×10^{2}	17×10^{2}
Hha I	0	0	0	0
	0.1	ND	1×10^{1}	2×10^{1}
	0.5	0	4×10^{1}	3×10^{1}
	5.0	0	8×10^{1}	12×10^{1}
Eco R1	0	0	0	0
	0.1	0	5×10^{1}	4×10^{1}
	0.5	0	9×10^{1}	ND
	5.0	0	8×10^{2}	6×10^{2}
Hpa I	0	0	0	0
	0.1	0	0	0
	0.5	0	0	0
	5.0	0	0	0
Hind III	0	0	0	0
	0.1	0	0	0
	0.5	0	0	0
	5.0	0	0	0

Legend to Table 1. Fifty microgram aliquots of $Ad2^{+}D1$ DNA were digested with each of several restriction endonucleases. At the end of the reaction the NaCl concentration was adjusted to 0.1 M and SDS and EDTA were added to final concentrations of 0.5% and 0.1 M respectively. The DNA fragments were purified by phenol extraction, concentrated by ethanol precipitation and dissolved in a small volume of Tris buffered saline. Confluent monolayers of CV-1 cells were infected with mixtures of fragments and SV40 tsa 30 DNA in the presence of DNA as described by Mertz and Berg (9). After twelve days incubation at 40.5°C the cells were stained with neutral red. Table shows the average number of plaques present on day 14. The experiment was performed in triplicate.

of Ad2^{+}D1 DNA by endonucleases Sma I, Bgl I, Hha I and Eco R1 resulted in the appearance of plaques with morphology typical of those caused by SV40. Neither Sma I, Hha I nor Eco R1 cleave within the SV40 sequences present in the Ad2^{+}D1 genome (see Figure 2): Bgl I cleaves the integrated SV40 sequences once, at position 67 (22) - close to the origin of DNA synthesis but not within the A gene. Approximately equal numbers of plaques were obtained in the presence of 0.1 μg and 0.5 μg of SV40 tsa 30 DNA - a result that was expected because preliminary experiments had shown 0.01 μg of SV40 DNA to be sufficient to saturate all the competent cells in a confluent monolayer 50 mm in diameter. From these experiments then it seems that fragments of Ad2^{+}D1 DNA are able to complement the growth of SV40 tsa 30 as long as they contain an intact copy of the SV40 A gene.

Isolation of SV40 adenovirus-2 hybrids:

Twenty-four plaques were picked from the complementation tests between SV40 tsa 30 DNA and fragments obtained by digestion of Ad2^{+}D1 DNA with endonucleases Hha I, Bgl I, Eco RI and Sma I. In all, then, ninety-six isolates were obtained and used to infect small (1 cm diameter) monolayers of CV-1 cells growing in plastic trays (Linbro Ltd., New Haven). After six days incubation at 40.5°C, when they were showing severe cytopathic effect, the cells together with their supernatant medium were frozen at -20°C, thawed and removed from the plates by aspiration. The resulting fluid was used to infect 5 monolayers of CV-1 cells, growing in 100 mm petri dishes. After three days incubation at 40.5°C the viral DNA was extracted by the method of Hirt (14) and purified by centrifugation to equilibrium in CsCl-ethidium bromide gradients as described earlier. About one microgram of each preparation of closed circular DNA was applied to a 1.0% agarose gel (20 x 20 cm) and subjected to electrophoresis for 20 hr. at 40V. The DNA was then denatured in situ, transferred to nitrocellulose sheets (18), and hybridized to ^{32}P labeled adenovirus 2 DNA as described (21). The presence of adenovirus 2 DNA sequences in the various isolates was then determined by autoradiography - a technique which reconstruction experiments have shown to be capable of detecting as little as 10^{-13} μg of DNA.

The results of an experiment in which sixteen isolates were assayed simultaneously is shown in Figure 3.

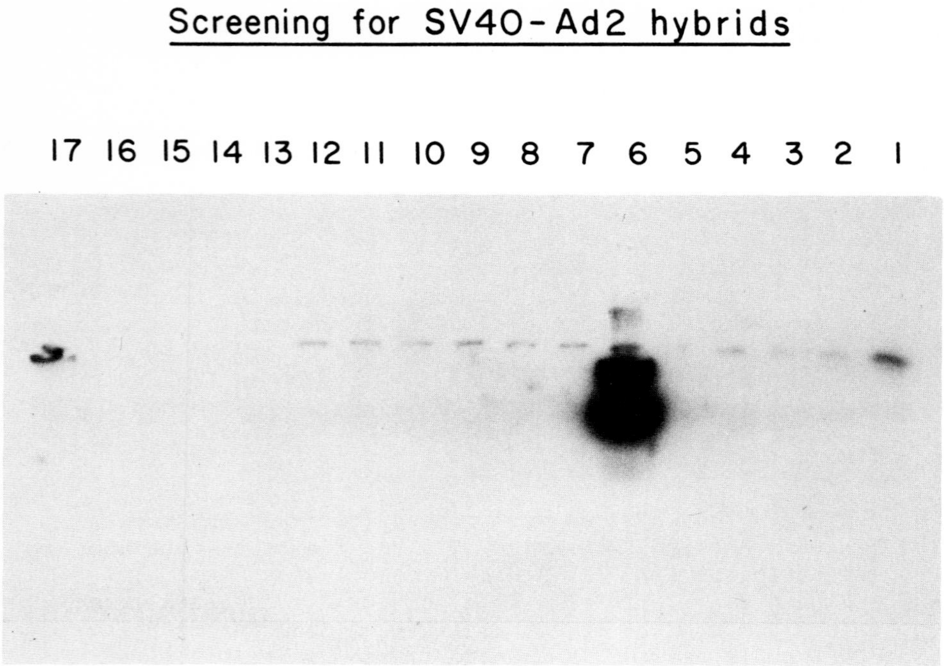

Figure 3. About 0.1 μg of closed circular DNA isolated from cells infected with sixteen putative SV40-adenovirus 2 hybrids were applied to an agarose gel and subjected to electrophoresis as described in the text. The bands of DNA were then denatured and transferred to a sheet of nitrocellulose (18) and hybridized (21) to about 10^7 cpm of ^{32}P labeled adenovirus 2 DNA (sp act 10^8 cpm/μg). After extensive washing the nitrocellulose sheet was exposed to X-ray films for 120 hours.

Slot 17 contains a small quantity of ^{32}P labeled component II SV40 DNA that served as a marker. Slot six shows two bands of hybridization to adenovirus 2 that correspond in mobility to components I and II of SV40 DNA. The DNA applied to the slot comes from an isolate called Sma 6.

Most of the isolates showed little or no homology to adenovirus DNA: One of them hybridized strongly and the position of the two bands corresponds to the location in the gel of the closed and relaxed circular forms of SV40 DNA. In all three viruses (Hha 7, Sma 2 and Bgl 6) were found whose DNAs hybridized well to adenovirus 2 DNA: stocks of these were prepared by two further serial passages in CV-1 cells. Four other isolates showed a lesser but still significant amount of homology to adenovirus 2 DNA: however no further work was carried out with them.

Analysis of the DNAs of Bgl 6, Hha 7 and Sma 2

Closed circular DNA was prepared from each of the three isolates which hybridize well to adenovirus 2 DNA, digested with restriction endonucleases and examined by gel electrophoresis and hybridization. The results obtained are shown in Figure 4.

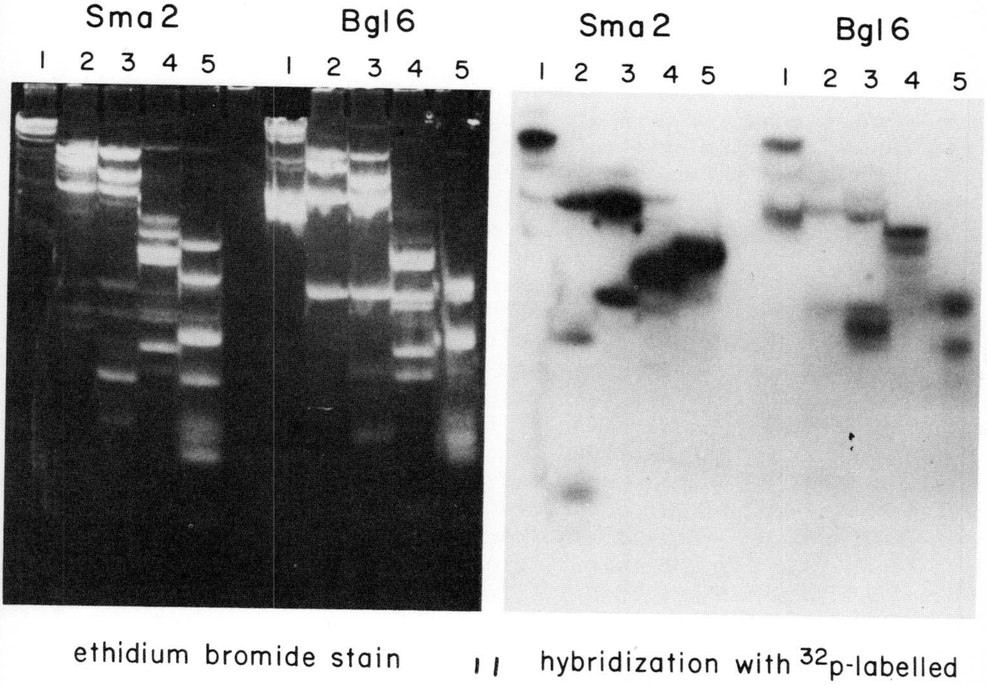

Figure 4. Analysis of the DNAs of Sma 2 and Bgl 6. About 1 μg of the DNAs of Sma 2 and Bgl 6 were digested with various restriction endonucleases and the resulting fragments were separated by electrophoresis through an agarose gel. The bands of DNA were stained with ethidium bromide and photographed (17) before they were denatured and transferred to a nitrocellulose sheet (18) and hybridized (21) to 5.10^6 cpm of ^{32}P-labeled adenovirus 2 DNA (sp. act 10^8 cpm/μg). After extensive washing the nitrocellulose sheet was exposed to X-ray film for 96 hours. Slot 1 contains DNA cleaved with Sma 1, slot 2 DNA cleaved with Eco R1, slot 3 DNA cleaved with Hha 1, slot 4 DNA cleaved with Hpa 1 and slot 5 DNA cleaved with Hind III.

The immediate impression is one of heterogeneity and it is at once obvious that the preparations of DNA must contain several different species. However the practised eye can discern amongst the complex pattern of fragments, familiar bands that are typical of digests of SV40 DNA. In all probability these are derived at least in part from the SV40 tsa 30 helper that is present in the stocks. New fragments of DNA, in abnormal molar quantities are found in digests of both Bgl 6 and Sma 2 DNAs (Figure 4): it is clear that the genomes present in these two viral stocks are not identical.

The fragments of DNA were transferred from the gel to a sheet of nitrocellulose and hybridized to ^{32}P-labeled adenovirus 2 DNA. It is clear that not all fragments of DNA made visible by ethidium bromide staining contain adenovirus sequences. For example only one fragment (approximately 3KB in size) obtained by cleavage of Sma 2 DNA with restriction endonuclease Hind III hybridized to adenovirus 2 DNA. A similar result was found when endonuclease Hpa I was used except that the molecular weight of the single DNA band which contained adenovirus sequences was approximately 2.8KB in length. However when Sma 2 DNA was digested with Hha I or Hae II - restriction endonucleases that cleave SV40 DNA very infrequently and adenovirus 2 DNA very often - several fragments were obtained that contained adenovirus sequences. The pattern of hybridization of fragments of Sma 2 DNA is different from that observed in the DNA of Bgl 6. In the latter case, no enzyme was found to yield a single fragment that contained all the adenoviral DNA sequences: and it often was difficult to identify the bands of DNA that hybridized to adenovirus DNA, so faintly did they stain with ethidium bromide. Apparently, in Bgl 6, adenovirus 2 sequences are contained within a species of DNA that is present in very low concentration. The same conclusion holds true for Hha 7 DNA, whose pattern of hybridization to adenovirus 2 DNA is shown in Figure 5. With endonuclease Eco R1, adenoviral sequences are found to occur in one major and two subsidiary, smaller DNA fragments: treatment of Hha 7 DNA with endonuclease Hae III produces several very small species of DNA whose faint hybridization with adenovirus 2 DNA can be seen only after prolonged exposure of the autoradiogram (Figure 5 B). Most interesting however is the pattern of hybridization found amongst the fragments obtained by digestion of Hha 7 DNA with endonuclease Hpa I. Five evenly-spaced bands are seen (Figure 5) whose ability to hybridize to adenovirus 2 DNA decreases with decreasing molecular weight. The bands differ in size from each other by about 200 base-pairs. This result must mean that Hha 7 contains a heterogeneous population of

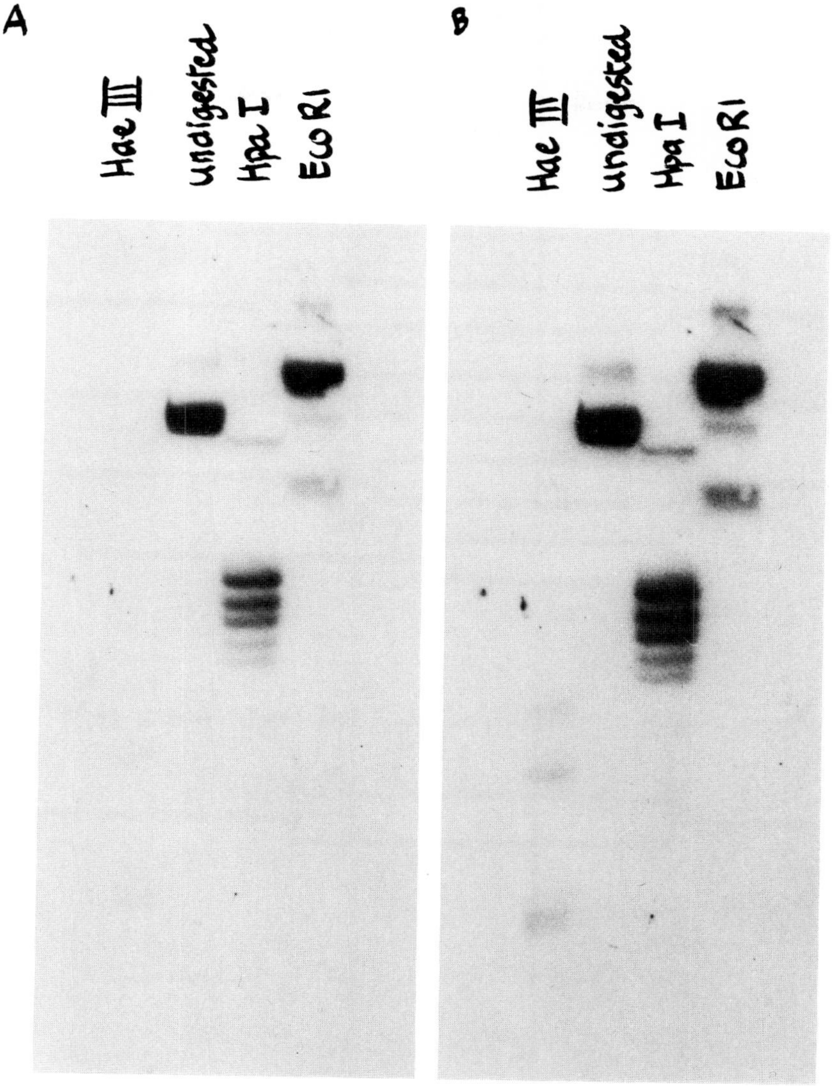

Figure 5. Distribution of adenovirus 2 sequences in DNA fragments of Hha 7.

About 0.3 μg of Hha 7 DNA was digested with restriction endonucleases, subjected to agarose gel electrophoresis, transferred to nitrocellulose filter and hybridized to ^{32}P-labeled adenovirus 2 DNA. The autoradiogram was exposed for 4 days (photograph A on the left) or 10 days (photograph B).

DNA molecules carrying adenovirus 2 DNA sequences which are distributed in one of two ways (see Figure 6). The first of these two arrangements seems by far the more likely.

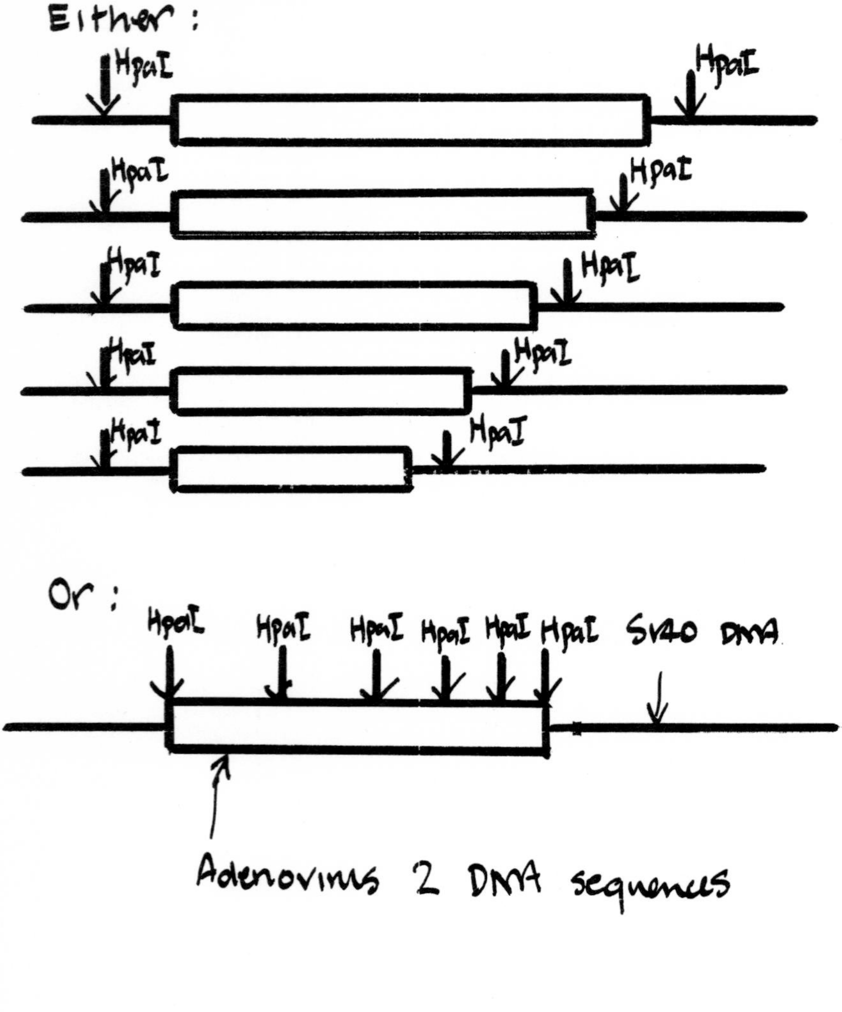

Figure 6. Possible arrangements of adenovirus 2 DNA sequences in the Hha 7 population.

It is obvious from all this that the isolates called Hha 7, Bgl 6 and Sma 2 do not contain a single substituted viral DNA species. They were therefore repurified by plaque isolation from monolayers of CV-1 cells infected and maintained at 40.5°C for 12 days. All three isolates formed plaques with two-hit kinetics. Several plaques of each isolate were picked and the production of working stocks and the process of isolation of closed circular DNA were repeated. The results of analyzing seven of the resulting isolates are shown in Figures 7 and 8. Obvious heterogeneity still is

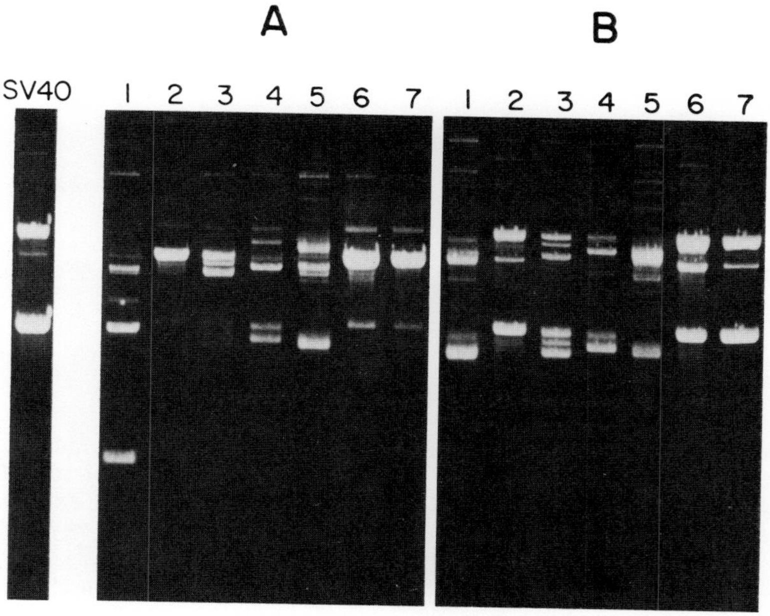

Figure 7. Analysis by agarose gel electrophoresis of subisolates of Hha 7, Sma 2 and Bgl 6.

About 1 μg of each isolate was applied to an agarose gel duplicate samples were treated with Eco R1. After electrophoresis the bands of DNA were stained with ethidium bromide and photographed (17). Photograph A shows the bands obtained after cleavage of the DNAs with Eco R1: photograph B shows the untreated DNAs. Slots 1, 3, 4 and 5 contain subisolates of Sma 2, slot 2 a subisolate of Hha 7 and slots 6 and 7 subisolates of Bgl 6 and 7.

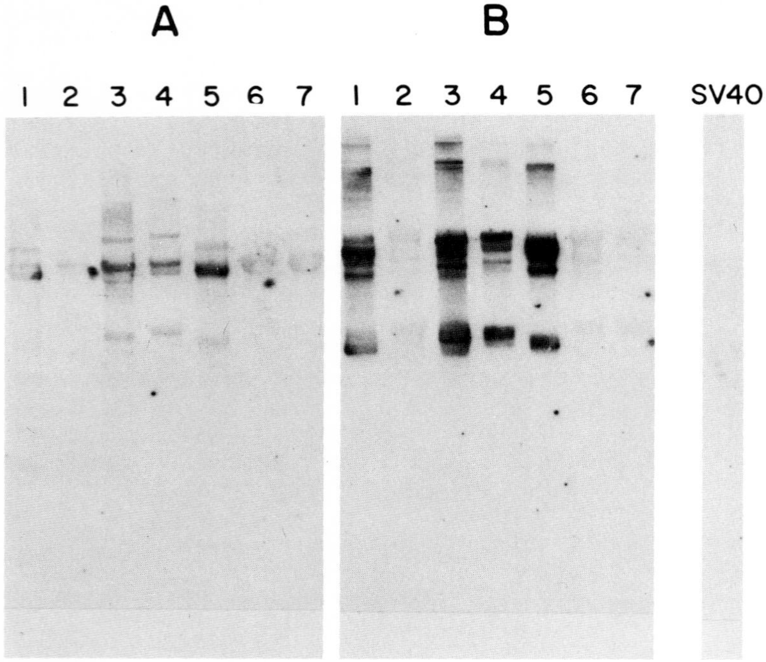

Figure 8. The DNAs shown in Figure 8 were transferred to a nitrocellulose sheet (18) and hybridized (21) to 2 x 10^6 cpm ^{32}P-labeled adenovirus 2 DNA. Autoradiographic exposure was for 4 days.

present, although there seems to have been significant simplification of the species of DNA in the preparations. Each isolate contains at least two sorts of DNA which can be either slightly smaller (Figure 7B; slots 1, 3, 4 and 5) or larger (slot 2) than standard-sized SV40 DNA. Some DNAs are resistant to cleavage by endonuclease Eco RI (Figure 8A; slots 4, 5, 6 and 7), while others seem to be almost completely sensitive to the enzyme (Figure 7A; slot 2): one isolate (slot 1) consists primarily of a species of DNA that contains two Eco RI cleavage sites.

The DNAs of all the subisolates hybridize with adenovirus 2 DNA, although there is wide variation in the amounts of adenoviral sequences detected (Figure 8). The greatest quantities are found in the DNAs of subisolates 1, 3, 4 and 5, all of which are derived from Sma 2: it is easy to see that in every case one major species and several minor species of DNA are present that hybridize with adenovirus

2 DNA. Subisolate 2 is derived from Hha 7, numbers 6 and 7 from Bgl 6. Again, heterogeneity is evident, and I conclude that repurification of the original isolates has improved the situation, but not eliminated the problem of variation within the hybrid viral stocks.

The origin of the adenovirus 2 DNA sequences in subisolates of Hha 7, Sma 2 and Bgl 6

To determine what part of the adenovirus 2 genome is present in each of the substituted genomes, hybridizations were carried out between unlabeled fragments of adenovirus 2 DNA, transferred directly from analytical agarose slab gels to nitrocellulose sheets, and the DNAs of subisolates 6, 4 and 2, which had been radiolabeled *in vitro* with ^{32}P (19, 20). The results are shown in Figures 9, 10 and 11. Knowing the map locations of the adenovirus 2 DNA fragments, it is a simple matter to work out the origin of the sequences that are present in the substituted genomes. Figures 9 and 10 show that Sma 2 and Bgl 6 contain indistinguishable sets of adenoviral sequences, which originate between positions 60 and 64 on the map of the adenovirus genome - a result that might have been predicted from the structure of the original starting material - $Ad2^{+}D1$ DNA (see Figure 2). However when the same experiment was carried out with DNA of Hha 7 (subisolate 2), a very surprising result was obtained (Figure 11): instead of those that flank the SV40 insertion in $Ad2^{+}D1$, Hha 7 (subisolate 2) carried sequences from a very distant part of the adenovirus genome which maps at the right hand end between map positions 94 and 98.

DISCUSSION

To summarize the data: defective viruses have been isolated whose closed circular genomes, ranging in size from 5-6KB, consist of covalently-joined SV40 and adenovirus 2 DNA sequences. Each defective virus stock contains a population of hybrid genomes that are different from those of the other stocks and heterogeneous in their own right. The detailed arrangement of the substituted genomes is not known, but in the simplest case (Sma 2), analyses by restriction endonuclease digestion and hybridization indicate that the adenovirus 2 DNA sequences are contiguous and have a maximum size of 2.8KB. Different hybrids contain sequences from different parts of the adenovirus 2 genome. It was not surprising to find heterogeneity in the original isolates of Hha 7, Sma 2 and Bgl 6. With the techniques currently available, exposure of cultured cells to a mixture of DNA

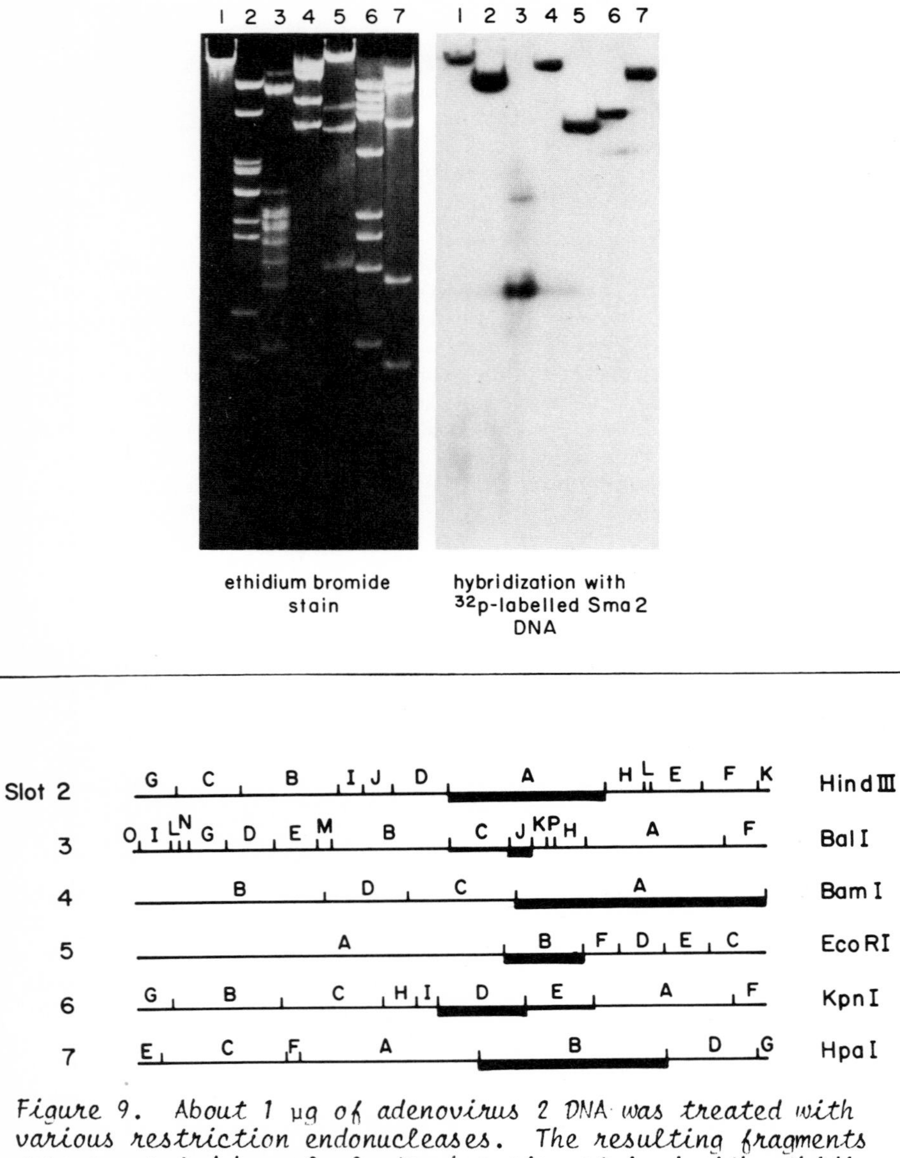

Figure 9. About 1 μg of adenovirus 2 DNA was treated with various restriction endonucleases. The resulting fragments were separated by gel electrophoresis, stained with ethidium bromide (17) and photographed and then transferred to a nitrocellulose (18) sheet for hybridization (21) to Sma 2 DNA ^{32}P-labeled in vitro (19, 20). Slot 1 contains uncut adenovirus 2 DNA. The other slots contain fragments cleaved by the enzymes listed in the lower part of the figure. Fragments that hybridize strongly to Sma 2 DNA are marked on the maps with thick lines, and those that hybridize weakly, with thinner lines.

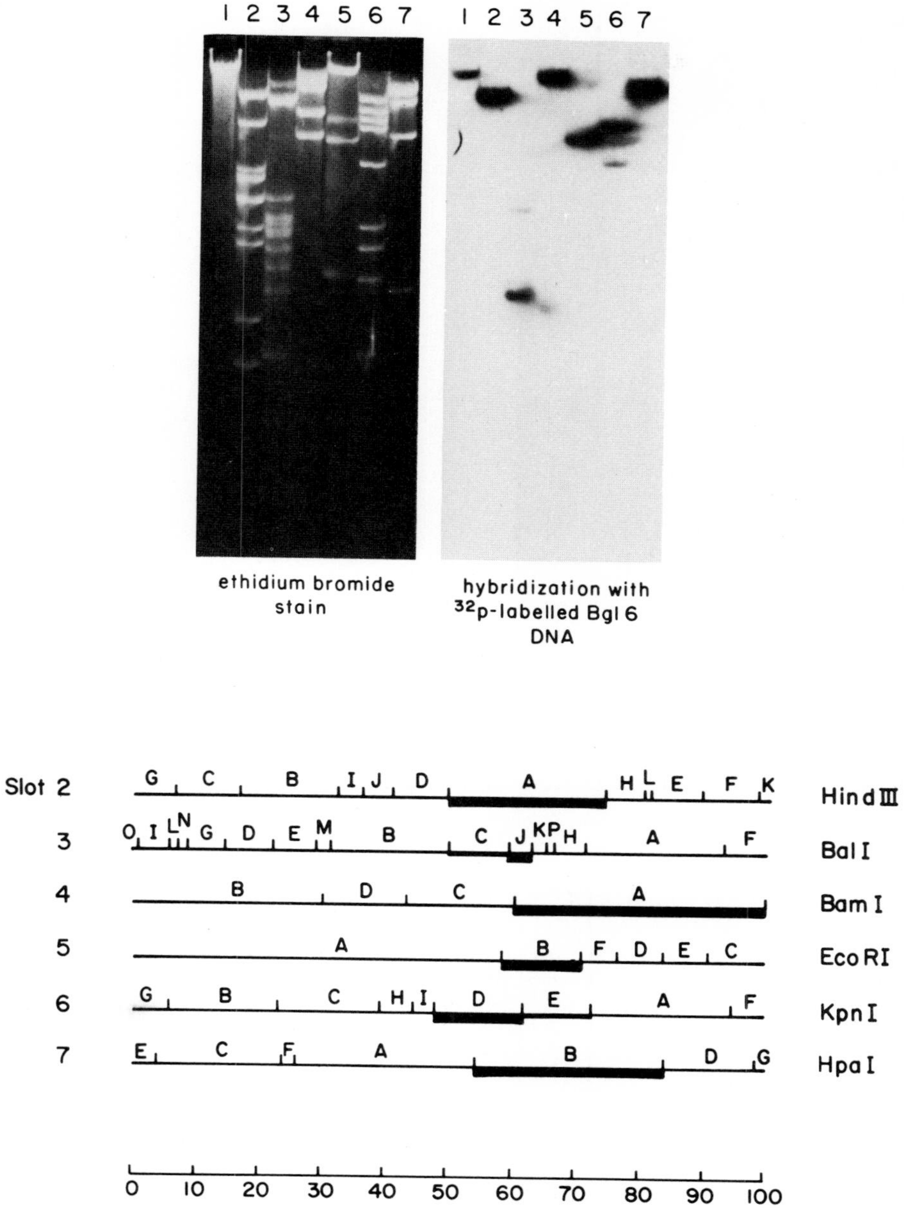

Figure 10. For experimental details see the legend to Figure 9.

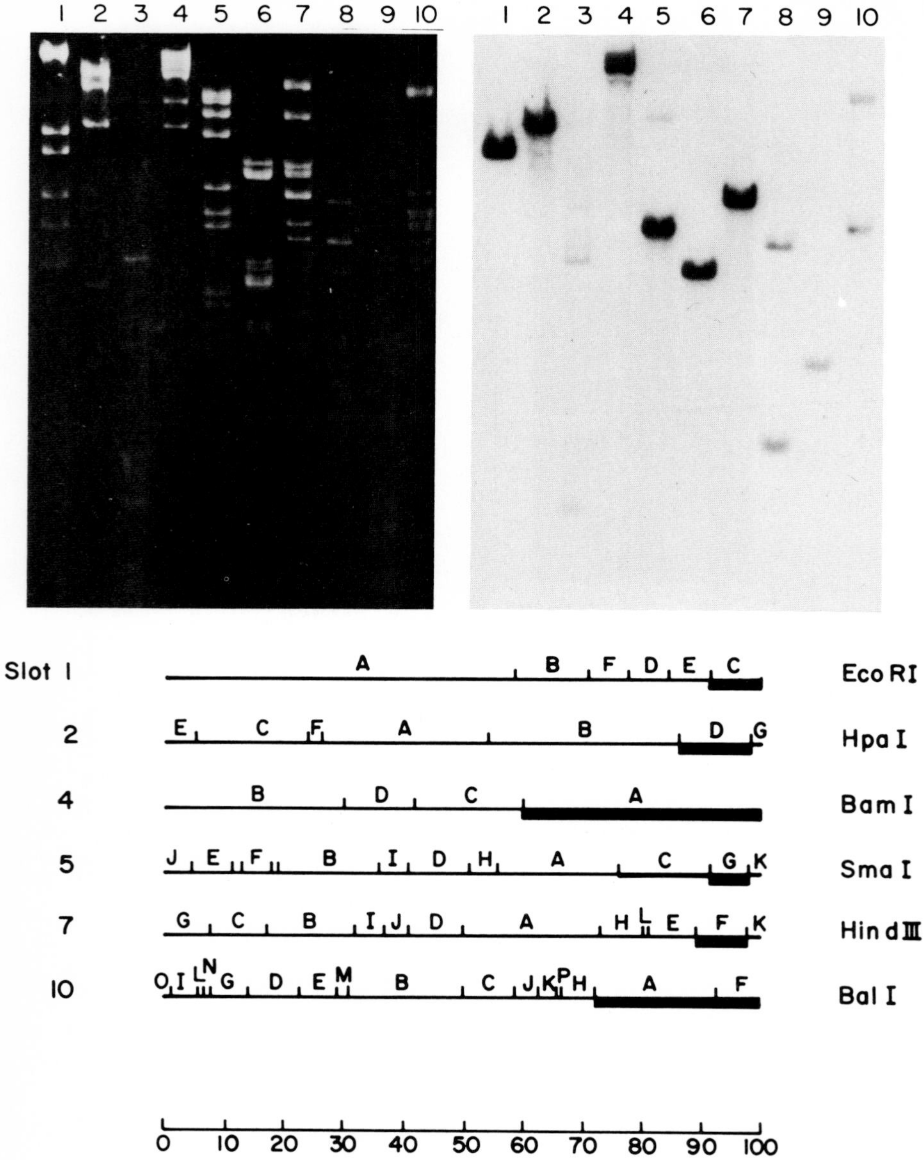

Figure 11. For experimental details see legend to Figure 9.

fragments will almost ensure that any infected cell will receive many kinds of molecules - probably in the form of a large clump. It seems entirely possible that this situation may promote recombination and result in the formation of several sorts of hybrid molecules. It is more disconcerting, however to find the variegation persisting through subsequent plaque purification. The causes of this heterogeneity are not understood and it is not known whether the SV40-adenovirus 2 hybrids are any more susceptible to amendment in vivo than are other sorts of substituted SV40 genomes. However it should be pointed out that Ganem et al (5), using gel electrophoresis were able to detect within the DNA sequences of SV40-λ hybrids, heterogeneity that appeared during passage of the substituted genomes in monkey cells. The hybridization techniques used here are considerably more sensitive and are capable to detecting rearranged molecules that are present at extremely low concentrations. It is obvious that the same techniques should be applied to other substituted SV40 genomes, both naturally-occurring and artificially generated.

How the SV40-adenovirus 2 hybrids were formed is not understood. In two cases (Bgl 6 and Sma 2) the cells were infected with intact SV40 tsa 30 DNA and an assorted mixture of fragments amongst which was a piece of DNA that contained the SV40 A gene and flanking adenovirus 2 DNA sequences. Presumably this linear fragment was converted to a covalently-closed molecule by a process of cell-mediated ring closure similar to that discussed by Lai and Nathans (23) and Mertz et al (24). The SV40 sequences presumably provide to the helper ts virus, a supply of the SV40 A gene product: in turn the helper virus donates capsid proteins which are used to pack the hybrid genome into virus particles.

The formation of Hha 7 is more mysterious. However the following series of events seems plausible. Endonuclease Hha I, by contrast to Bgl 1 and Sma 1 cleaves adenovirus 2 DNA into very small pieces. In fact it is known (Sambrook, unpublished) that the distance from the SV40 insertion in $Ad2^{+}D1$ to the nearest Hha I cleavage sites in both directions is less than 100 base-pairs. Even if this DNA could circularize without destroying the SV40 A gene, the resulting molecule would be too short to be packed efficiently into SV40 virions (5). The longer genome of Hha 7 must have been formed by recombination between the SV40-containing fragment of $Ad2^{+}D1$ and a piece of DNA that, although derived from a distant part of the adenovirus 2 genome, happened to be conveniently adjacent in the infected cell. This result once again points up the bizarre processes of recombination that occur in cells infected by SV40: it seems to be a commonplace that DNAs which share little or no base sequence homology, are genetically remote from one another, show little or no organizational similarities and on all these grounds could be expected not to interact, nevertheless manage to recombine to form hybrids. The substituted genomes described in this paper can therefore be added to a list of products of promiscuous recombination that already includes integration of adenovirus SV40 hybrid viruses, and rearrangements and substitution within the SV40 genome itself.

FOOTNOTE

The use of animal viruses as vectors is a combustible topic and the growth of recombinant DNAs made by *in vitro* ligation of adenovirus and SV40 DNAs is prohibited unless carried out under P4 conditions (25). By relying upon cellular recombination enzymes I have not crossed any of the guidelines and yet been able to generate *in vivo* the very hybrids whose creation is proscribed by N.I.H.

Because of this paradox I have decided not to continue work with these hybrids nor to distribute them. What is interesting about them is not their structure nor even their expression, but the mechanism by which they were generated: and there are ways of analyzing this process without compromising the N.I.H. guidelines.

ACKNOWLEDGMENTS

This work was supported by funds supplied by the National Cancer Institute. I am indebted to Denise Galloway for numerous gifts of nick-translated DNAs. I thank Terri Grodzicker, Michael Botchan and Denise Galloway for stimulating discussions and Maxine Singer and Paul Berg for constructive advice about ethical matters.

REFERENCES

(1) S. Lavi and E. Winocour. J. Virol. 9 (1972) 309.

(2) S. Rosenblatt, S. Lavi, M.F. Singer and E. Winocour. J. Virol. 12 (1973) 501.

(3) N. Frenkel, S. Lavi and E. Winocour. Virology 60 (1974) 9.

(4) S. Segal, M. Garner, M.F. Singer and M. Rosenberg, Cell 9 (1976) 247.

(5) D. Ganem, A.L. Nussbaum, D. Davoli and G.C. Fareed. Cell 7 (1976) 349.

(6) E. Lukanidin, J. Hassell and J. Sambrook. (1977) in preparation.

(7) G. Fey, E. Lukanidin and J. Sambrook. (1977) in preparation.

(8) A.S. Rabson, G.T. O'Conor, I.K. Berezesky and F.J. Paul. Proc. Soc. Exp. Biol. Med. 116 (1964) 187.

(9) J. Mertz and P. Berg. Virology 62 (1974) 112.

(10) P. Tegtmeyer. J. Virol. 10 (1972) 591.

(11) T. Grodzicker, C.W. Anderson, P.A. Sharp and J. Sambrook. J. Virol. 13 (1974) 1237.

(12) A.M. Lewis, Jr. and W.P. Rowe. J. Virol. 5 (1970) 413.

(13) U. Pettersson and J. Sambrook. J. Mol. Biol. 73 (1973) 125.

(14) B. Hirt. J. Mol. Biol. 26 (1967) 365.

(15) J. Sambrook, J. Williams, P.A. Sharp and T. Grodzicker. J. Mol. Biol. 97 (1975) 369.

(16) W. Sugden, B. DeTroy, R.J. Roberts and J. Sambrook. Anal. Biochem. 68 (1975) 36.

(17) P.A. Sharp, W. Sugden and J. Sambrook. Biochemistry 12 (1973) 3055.

(18) E. Southern. J. Mol. Biol. 98 (1975) 503.

(19) P.W.J. Rigby, D. Rhodes, M. Dieckmann and P. Berg. in preparation.

(20) T. Maniatis, S.G. Kee, A. Efstradiatos and F.C. Kafatos. Cell 8 (1976) 163.

(21) M. Botchan, W.C. Topp and J. Sambrook. Cell 9 (1976) 269.

(22) S. Zain. personal communication.

(23) C-J. Lai and D. Nathans. Cold Spring Harbor Laboratory Symposium on Quantitative Biology 39 (1974) 53.

(24) J.E. Mertz, J. Carbon, M. Herzberg, R.W. Davis and P. Berg. Cold Spring Harbor Laboratory Symposium on Quantitative Biology 39 (1974) 69.

(25) Federal Register. Part II: Department of Health, Education and Welfare, National Institutes of Health. Recombinant DNA Research Guidelines. Wednesday, July 7, 1976.

THE GENETIC MAP OF ROUS SARCOMA VIRUS

Peter H. Duesberg, Lu-Hai Wang, Pamela Mellon,
William S. Mason[1] and Peter K. Vogt[2]

Department of Molecular Biology and Virus Laboratory,
University of California, Berkeley, Ca. 94720
[1]Institute for Cancer Research, Philadelphia, Pa. 19111
[2]Department of Microbiology, University of Southern
California, Los Angeles, Ca. 90033

ABSTRACT

The four genetic elements of Rous sarcoma virus (RSV), gag coding for internal group-specific proteins, pol coding for the viral DNA polymerase, env coding for the viral envelope glycoprotein, and src responsible for sarcoma formation, have been mapped on the viral RNA genome of 10,000 nucleotides by a chemical method. By this method we order 20 to 30 RNase T_1-resistant oligonucleotides relative to the 3'poly(A)-end of the RNA to derive an oligonucleotide map. Oligonucleotides are functionally identified by correlating their appearance in viral RNA with genetic markers.

By comparing the distributions of genetic markers with the oligonucleotides of RSV having a temperature sensitive (ts) polymerase, termed tsLA 337 or 335, a leukosis virus RAV-6, and twenty of their recombinant progeny, pol-, env-, and src-specific oligonucleotides were identified. Gag-oligonucleotides were identified by analyzing the RNAs of a RSV with a ts gag gene, termed LA 3342 and of four recombinants generated by crossing-over with RAV-6. In accord with previous results, recombinants contained (i) src-oligonucleotides between the poly(A)-end and 2000 nucleotides, (ii) env-oligonucleotides between 2500 and 5000 nucleotides, and (iii) pol-oligonucleotides between 6000 and 8000 nucleotides from the poly(A)-end of the RNA. Pol-oligonucleotides were shared by all recombinants with their parents and were identified by exclusion of other map segments which failed to segregate with pol-markers. (iv) The 5' segment of recombinant RNAs, between 8000 and 10,000 nucleotides from the poly(A), contained two or three oligonucleotides that segregated with their parental gag gene marker. The gene order 5'gag-pol-env-src-poly(A) was deduced.

INTRODUCTION

Nondefective Rous sarcoma virus (RSV) contains four genetic elements: gag, coding for the internal structural proteins termed group-specific antigens; pol, coding for the viral DNA polymerase; env, coding for the envelope glycoprotein, and src coding for sarcoma formation (1,2). The viral genome is an RNA of 10,000 nucleotides (3,4,5), with a poly(A) stretch at the 3' end (6,7). We have developed a chemical method to map these genetic elements on the viral RNA. To apply this method, three conditions have to be met: RNA segments corresponding to a genetic element have to be (a) chemically and (b) functionally identified and (c) have to be located on the viral RNA. We identify RNA segments by their large RNase-T_1-resistant oligonucleotides which altogether represent only 5% of the RNA (4,5), and are detected by two-dimensional electrophoresis-chromatography, termed fingerprinting (8). The location of a given oligonucleotide, and of the RNA segment represented by it, relative to the 3' poly(A)-terminus of the RNA is then deduced from the length of the smallest poly(A)-tagged RNA fragment from which it can be isolated. The resulting order of all large oligonucleotides is termed an oligonucleotide map (7,9,2,10,11,12). The genetic functions of RNA segments represented by distinct oligonucleotides are identified by three procedures: (i) Deletion mapping: Oligonucleotides present in nondefective viruses but absent from corresponding deletion mutants are equated with the genetic function of the wild type that is lacking in the deletion (2,7,8,10); (ii) Recombinant mapping: Viral recombinants contain a specific selection of oligonucleotides from their parents (4,5). Therefore the genetic function of a given oligonucleotide can be identified if it always segregates with a parental gene marker in recombinants (11, 12,13,14,15); (iii) Functionally identifying conserved oligonucleotides: this empirical procedure relies on the observation that certain viral oligonucleotides are highly conserved in all avian tumor viruses and if functionally identified can be used to identify genes in a family of related viruses (9,12,13,14,15). Applying the above methods and considering evidence obtained from in vitro translation of viral RNA (16) and genetic linkage data from three factor-crosses among different viruses (13,17,18,19), we have recently proposed that the complete genetic map of RSV is 5'gag-pol-env-src-poly(A) (13,14).

This proposal was based on unambiguous identifications of src- and env-specific oligonucleotides by procedures (i) and (ii). Pol-specific oligonucleotides were identified as some or all of a cluster of four conserved oligonucleotides present in twenty recombinants and both of their parents, because the oligonucleotides of all other map segments of these recombinants segregated independently of their pol-phenotype and therefore could be excluded (13,14). The gag gene was identified indirectly by excluding src-, env-, and pol-specific map segments, and by a good, but not complete, correlation between an electrophoretic marker of the viral p27 protein, a major gag gene product, with the segregation of a cluster of four oligonucleotides in the 5' segment of recombinant RNAs (14). However, a p27-specific oligonucleotide could not be identified in the recombinants investigated, because of electrophoretic variations found in the p27 proteins of some recombinants which did not covary with any specific oligonucleotide. This is due to the expected limit of the method. The 20 to 30 RNase-T1-resistant oligonucleotides in each oligonucleotide map represent only 5% of the viral RNA, because 95% of the RNA is digested by RNase-T_1 to oligonucleotides which are too small for detection by fingerprinting. Since P27 is coded by only 8% of the RNA, it is possible that the coding region for p27 may not contain a large oligonucleotide, and probable that variations among the p27 proteins of recombinants would not be reflected by an altered oligonucleotide.

The purpose of this paper is to identify gag- and pol-specific oligonucleotides of recombinant virus RNAs. Two classes of recombinants were analyzed here. One had been selected from crosses between Prague RSV of subgroup C, having a temperature-sensitive polymerase, termed tsLA 337 and 335 (21), and leukosis virus RAV-6 of subgroup B which lacks a src gene. Thus, differential parental markers of the pol, env, and src genes of each virus were known in this class of recombinants, while the parental origin of their gag gene was unknown. Part of these studies has already been described (13, 14). Another class of recombinants was selected from crosses between B77-RSV of subgroup C, having a temperature-sensitive gag gene, termed tsLA 3342 (17,33), and leukosis virus RAV-6 of subgroup B. The gag-lesion of tsLA 3342 renders virus particles, made at the nonpermissive temperature, noninfectious by interfering with the normal cleavage process that converts the primary gag gene-product pr76 (20) into smaller structural proteins of the virus (33). Thus differential parental markers of the gag, env, and src genes were known for this class of recombinants, while the parental origin of their pol gene was unknown.

Since it is not known where within the gag or pol genes the lesions of tsLA 3342 or of tsLA 337 and 335 are located, and whether these lesions affect one or several sites of each gene, we cannot predict how many nucleotides must be exchanged by crossing over to restore the normal phenotype. Crosses exchanging only a few nucleotides are not likely to be detected by our method, because they would probably generate wild-type recombinants which would have the same fingerprint pattern as that of the ts parent.

In contrast, crosses in which the ts-lesion is exchanged with much or all of the respective gene should be detectable by the exchange of several oligonucleotides adjacent to the ts-lesion. Since gag comprises about 23% (16,20) and pol about 30% (23,24) of the viral RNA and probably an equivalent share of the large oligonucleotides, several oligonucleotides would be expected to segregate closely with ts markers of these genes. Nevertheless many recombinants must be analyzed in order to identify such oligonucleotides. Obviously the more oligonucleotides that are already correlated with known genetic markers, the easier it is to identify the remaining oligonucleotides of a recombinant. Consequently we are focusing our search for gag and pol oligonucleotides on the 5' half of the oligonucleotide map, because previous results have indicated that src and env genes take up most of the 3' half of the RNA (9,12). The analyses described here confirm that pol-specific oligonucleotides map between 6000 and 8000 nucleotides, and indicate that gag-specific oligonucleotides map between 8000 nucleotides and the 5' end of the RNA, confirming the gene order 5'gag-pol-env-src-poly(A).

MATERIALS AND METHODS

All virus strains (13,14), and procedures for preparation of viral (^{32}P) RNA, fingerprinting of RNase T1-resistant oligonucleotides and autoradiography have been described in detail (7,9). Recombinant viruses tsLA 337/335 X RAV-6 have been described previously (13,14,21) and tsLA 3342 X RAV-6 recombinants have been prepared for this investigation as described by Vogt (22).

RESULTS

The Oligonucleotide Maps of tsLA 337/335, of Leukosis Virus RAV-6, and of 20 tsLA337/335 X RAV-6 Recombinants. The RNase-T_1-resistant oligonucleotides of tsLA 337, tsLA 335, RAV-6, and 20 tsLA 337/335 X RAV-6 recombinants, some of which are shown in ref.(13), were resolved by two-dimensional electrophoresis-chromatography, termed fingerprinting (Fig. 1). The fingerprint patterns of the parental viruses tsLA 337 (Fig. 1), and tsLA 335 (not shown), were identical. Prague RSV tsLA 337/335-specific oligonucleotides were designated as P-numbers and RAV-6-specific oligonucleotides as R-numbers (Table 1, Fig. 1 and 2). Oligonucleotides shared by both viruses were given unlettered numbers. Six of these common oligonucleotides are highly conserved with regard to composition and oligonucleotide map location (Fig. 2) in all avian tumor viruses analyzed to

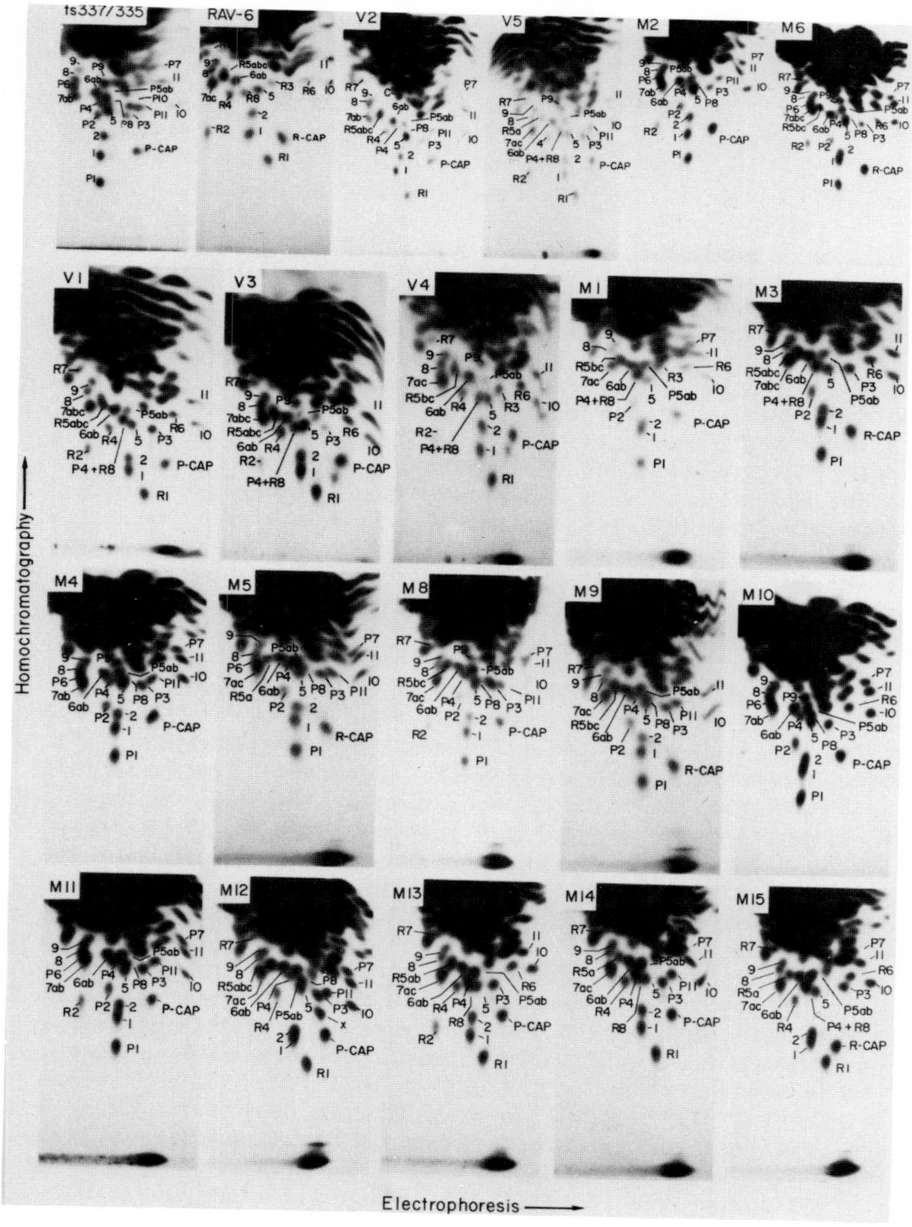

Fig. 1. RNase T₁-resistant oligonucleotides of the 60-70S (32P)RNAs of Prague RSV tsLA 337 (same as 335, not shown), leukosis virus RAV-6, and twenty tsLA 337/335 X RAV-6 recombinants, identified by V- and M-numbers, after electro-

phoresis and homochromatography (fingerprinting). Preparation of (^{32}P)RNA, its digestion with RNase T_1, fingerprinting, and autoradiography have all been described in detail (7,8,9). Fingerprints derived from RNA which had been alkali fragmented and from which poly(A)-tagged fragments had been removed, lack the poly(A)-spot seen in the lower right corner if unfractionated RNA is analyzed.

date (12). Analyses of the RNase A-resistant fragments of the large T_1-resistant oligonucleotides of each virus strain are reported in Table 1. In all previous analyses the fingerprints of recombinants were found to be mosaics of parental oligonucleotides (4,5,12). Consequently we have designated recombinant oligonucleotides with the symbols of their parental counterparts (Table 1, Fig. 2). However recombinant M12 contained one oligonucleotide termed x, (Fig. 1, Table 1), mapping 3000 nucleotides from the poly(A) end of its RNA which was not shared with either parent. This oligonucleotide was presumably generated by crossing-over.

Oligonucleotide maps of tsLA 337/335 and RAV-6 (Fig. 2) were derived by, determining from discrete sizes of poly(A)-tagged RNA fragments, the distance between each oligonucleotide and the poly(A) end of the viral RNA. Fingerprint patterns of these poly(A)-tagged fragments have been published (13). Some oligonucleotides shared by both parental viruses are connected by single horizontal lines and oligonucleotides highly conserved among all avian tumor viruses (12) are connected by double lines. Oligonucleotide maps of twenty tsLA 337/335 X RAV-6 recombinants are shown on the same figure. The oligonucleotide maps of ten recombinants (V1-V5, M2, M4, M6, M10, M12) were derived directly from fingerprints of poly(A)-tagged RNA fragments (13). The oligonucleotide maps of the remaining recombinants (M1, M3, M5, M7-9, M11, M13-15) were drawn by assigning for each recombinant oligonucleotide a map location which is equivalent to its parental counterpart. This is in accord with previous evidence that rearrangements and permutations do not occur during recombination and that all avian tumor viruses share highly conserved oligonucleotides at equivalent map locations (9,12,13,14). The presence of a parental oligonucleotide is indicated by + and the absence by - in three different columns at the appropriate map location for each oligonucleotide of a recombinant (Fig. 2). The right column records tsLA 337/335-specific oligonucleotides (P-numbers), the middle column records oligonucleotides shared by both parents (unlettered numbers), and the left column records RAV-6-specific oligonucleotides (R-numbers). The stippled area connecting recombinant oligonucleotides over two columns is an interpretation of our data. We assume that a

Table 1. RNase-T1-resistant oligonucleotides* of Prague RSV tsLA 337/335, RSV B77 tsLA 3342, RAV-6, tsLA 337/335 X RAV-6, and tsLA 3342 X RAV-6 recombinants

ts337	ts3342	RAV-6	RNase A digestion products†
C	C	C	G, (AC), (AU), (AAU), (AAAC)
1	1	1	4U, 8C, 2(AC), (AU), (AAC), (AAG)
2	2	2	5U, 8C, G, (AC), (AU), (AAAC)
5	5	5	5U, 6C, (AAG), (AAAAN)
6ab	6ab	6ab	4U, 6C, 2G, 4(AC), 2(AU), (AAC), (AAAAAN)
7b	7b		U, 3C, G, (AC), (AU), (AAAAAAN)
7a	7a	7a	6C, 2(AC), (AU), (AG), (AAC)
		7c	5C, 2(AC), (AU), (AG), (AAU), (AAAAAN)
8	8	8	U, 6C, 3(AC), (AAG)
9	9	9	3C, (AU), (AG), (AAAAAAN)
10	10	10	7U, 2C, G, 4(AU), (AAU)
11	11	11	9U, 6C, G, (AU)
P1	B1		7U, 8C, G, 3(AC), (AAC)
P2	B2		3U, 7C, G, 3(AC), (AU)
P3	B3		4U, 3C, G, (AC), 2(AU), (AAC), (AAU)
P4	B4		4U, 5C, G, 2(AC), (AU), (AAAC)
P5ab	B5ab		6U, 8C, G, 4(AC), 2(AU), (AG)
P6	B6		U, 7C, G, 3(AC), (AAAAN)
P7			5U, 3C, G, (AU), (AAU)
P8	B8		4U, 4C, (AC), (AAG), (AAAAN)
P9	B9		4U, 4C, (AC), (AAAG)
P10			5U, 3C, 2(AC), 4(AU), (AG)
P11			5U, 3C, G, (AC), 2(AU) (AAAAN)
	B12		U, 5C, G, 2(AC), (AU), (AAC)
P-CAP	B-CAP‡		4U, 4C, G, 3(AC), 3(AU), CAP
		R1	5U, 8C, G, (AC), 2(AU), (AAC), 2(AAU)
		R2	2C, G, 2(AC), 2(AAAAN)
		R3	3U, 3C, (AG), (AAU), (AAAU)
		R4	2U, 7C, G, 2(AC), (AU), (AAC)
		R5abc	5U, 12C, 3G, 6(AC), 2(AU), 1-2(AAC), 2(AAAN)
		R6	6U, 3C, G, 2-3(AU), (AAAAN)
		R7	3C, 0.5(AC), 2(AAC), (AAG)
		R8	5-6U, 6C, (AAG), (AAAAN)
		R-CAP‡	4U, 6C, G, 3(AC), 2(AU), CAP
x of M12§			5U, 3C, (AC), 0.5(AU), (AAC), 2(AAU), (AAG)

Table 1.

**Numbers, capital letters, or number-capital letter combinations refer to oligonucleotides of tsLA 337/335, tsLA 3342, RAV-6 (Figs. 1 & 3), or tsLA 337/335 X RAV-6 and tsLA 3342 X RAV-6 recombinants (see also text). Oligonucleotide spots containing more than one G residue are identified by lower-case letters to indicate the number of unresolved or partially resolved oligonucleotides overlapping in such a spot.*

†RNase-A-resistant fragments of RNase-T_1-resistant oligonucleotides were determined as described (4,5). Residue symbols in parentheses indicate nucleotide sequences.

‡This oligonucleotide was shown to contain m^7GpppG^mpCp, the 5'-terminal sequence of tumor virus RNA (25,26).

§Oligonucleotide x of recombinant M12 had no parental counterpart (see text).

given recombinant map segment was inherited from tsLA 337/335 if the stippled area connects + symbols of the middle and right column and that segments were derived from RAV-6 if + symbols of the middle and left column are connected. Switches would indicate a minimum number of cross-over points ranging between 2 and 5.

Comparing oligonucleotide map segments with the known genetic functions of the parents and their recombinants, we make the following seven deductions.

The Map Locations of src, env, and pol in Recombinants. (i) All recombinant RNAs shared between poly(A) and nucleotide 2000 three to four oligonucleotides (P3, P4, P5ab) with the tsLA 337/335 parent. Since all recombinants analyzed here were selected for the src gene of tsLA 337/335 and since RAV-6 lacks a src gene (leukosis viruses lack src (2,7,8,10)), it follows that the src gene for these recombinants resides in this map segment. This is consistent with previous analyses, which have mapped the src gene of Prague RSV-C, the wild-type of tsLA 337/335, in this position (7), and additional results which have indicated that the src oligonucleotides are highly conserved among all avian sarcoma viruses (9.12,13,14). In addition all recombinants shared with both of their parents and all other avian tumor viruses oligonucleotide C at the poly(A) end. (ii) Three conserved oligonucleotides (6ab, 7a) were found between 1800 and 2500 nucleotides in all recombinants and both parents. These oligonucleotides could belong to any viral gene except the src gene, which is absent from RAV-6. However, this map segment is probably too small to re-

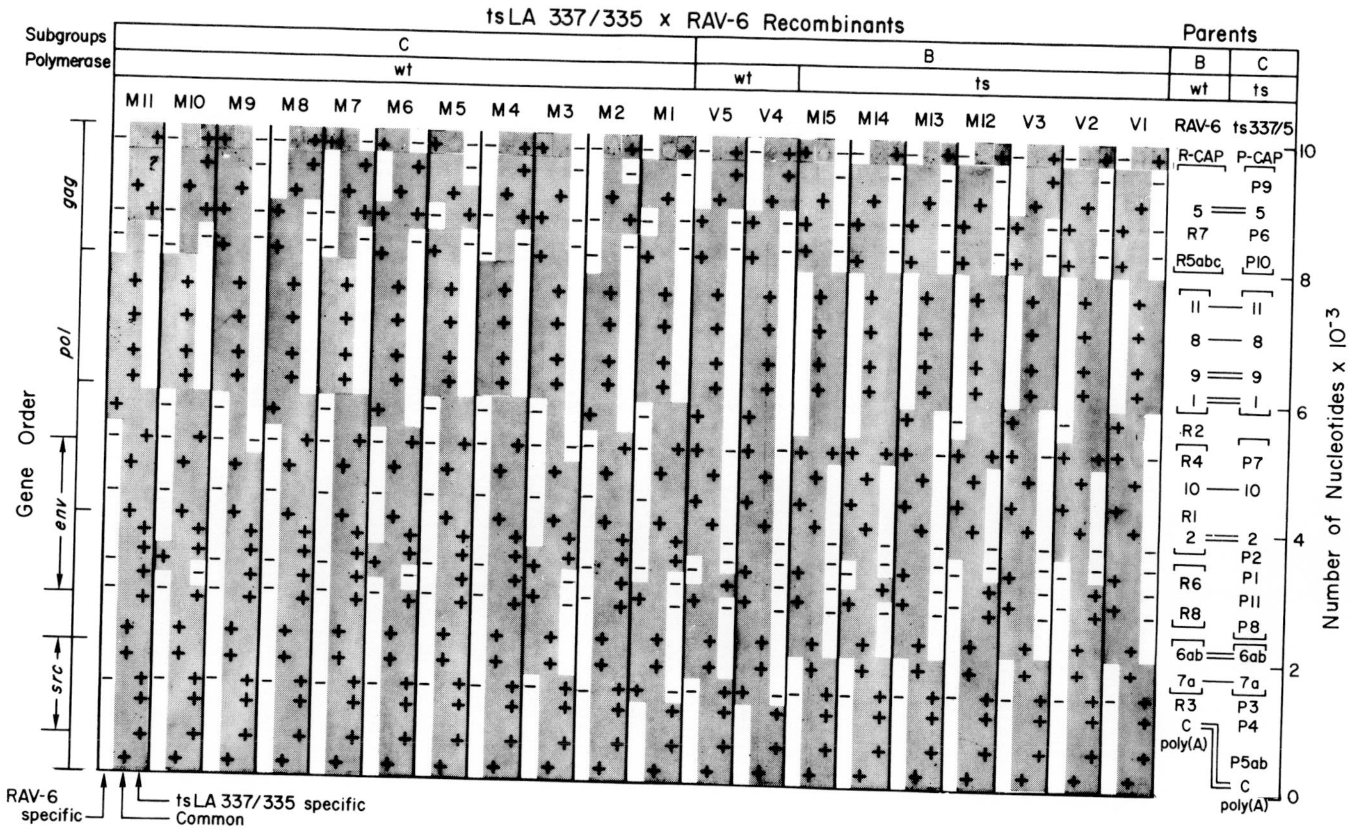

Fig. 2. Oligonucleotide maps of tsLA 337/335, RAV-6, and 20 tsLA 337/335 X RAV-6 recombinants. RNase-T1-resistant oligonucleotides are numbered as described for Fig. 1 and Table 1 and have

been ordered linearly on the basis of their distance from the 3'poly(A) end of the viral RNA (7,9). The right ordinate shows the size of viral RNA in nucleotides (4,5). The relative location of oligonucleotides within brackets is uncertain. The oligonucleotide map of RAV-6 is 15% shorter than that of tsLA 337/335 because RAV-6 lacks the src gene (see text). Horizontal lines connect oligonucleotides shared by both viruses at equivalent map locations. Double lines connect oligonucleotides highly conserved among avian tumor viruses (9,12). The left ordinate indicates the most probable gene order of avian tumor virus RNA derived from these and other data (text and refs. 7,8,10-15). The upper abscissa indicates the env- and pol-phenotypes of each virus. B and C are viral subgroups defined by the env gene. V- and M-numbers are symbols for specific recombinants (see Fig. 1). The oligonucleotide maps of the 20 recombinant RNAs show, in three different columns, the presence (+) or the absence (-) of a given parental oligonucleotide at the map location equivalent to its parental counterpart. The right column records tsLA 337/335-specific oligonucleotides, the left column RAV-6 specific oligonucleotides, and the middle column oligonucleotides shared with both parents. The stippled area connects all adjacent recombinant oligonucleotides between two columns as they might have been inherited from RAV-6 (left and middle column) or from tsLA 337/335 (right and middle).

present gag, which codes for a protein of 76,000 daltons (20), corresponding to 2300 nucleotides, or pol, which codes for a protein of 100,000 daltons (23,24), corresponding to 3000 nucleotides. Hence it is not likely an integral part of these genes (see below). Instead it may be a conserved part of the env gene which maps near src between 2500 and 5000 nucleotides (2,9,11-15), or it may represent a yet undefined genetic function of the virus. (iii) The map segments between 2500 and 5000 nucleotides are known to be part of the env gene (2,9,11-15). As expected (2,12), all recombinants selected for subgroup B env gene from RAV-6 had RAV-6-specific oligonucleotides and all recombinants selected for the subgroup C env from tsLA 337/335 had tsLA 337/335-specific oligonucleotides in this segment (Fig. 2). Moreover, both parents and all recombinants shared the highly conserved env-specific oligonucleotide (no. 2) in this map segment. (iv) Between 5000 and 6000 nucleotides recombinants either contained or lacked oligonucleotide R2 or P7 irrespective of their selected biological markers. Since these oligonucleotides did not segregate with a known genetic marker, their function cannot be determined. (v) All recombinants shared with both parents a map segment between

6000 and 8000 nucleotides defined by four conserved oligonucleotides. Two of these are highly conserved among all avian tumor viruses (12). Since no tsLA 337/335-specific oligonucleotide map segment was found that segregated with all ts pol recombinants and no RAV-6 map segment was found that segregated with all wild-type pol recombinants (see below), we concluded that the ts pol lesion of LA 337/335 is associated with the RNA segment identified by these conserved oligonucleotides.

Oligonucleotides near the 5' End of Recombinant RNAs Fail to Segregate with src, env, and pol. (vi) The 5'-terminal map segment, between 8000 and 10,000 nucleotides but not including the 5'-terminal CAP oligonucleotide (see below), was defined in some recombinants (V1, V2, M3, M9, M12-15) by RAV-6-specific oligonucleotides R7, R5abc, in others (M2, M4, M7, M10, M11) by tsLA 337/335 oligonucleotides P6 and P9, and in still others (V3, V4, V5, M5, M6, M8) by a combination of RAV-6 and tsLA 337/335 oligonucleotides (Fig. 2). The presence of three RAV-6 and two tsLA 337/335 oligonucleotides in recombinant M6 could be due to unequal crossing-over (4,5), or could signal a mixture of two viruses. Since the oligonucleotides of this map segment segregated independently of the three known viral gene markers of our recombinants, it would appear by exclusion that gag might be associated with this map segment. This suggestion is compatible with a good, but not complete correlation between an electrophoretic marker of p27 and oligonucleotides of the 5' RNA segment of these recombinants described previously (14), and is directly confirmed in experiments described below.

The 5'-Terminal Oligonucleotides. (vii) The 5'-terminal oligonucleotides, termed CAP (25,26), were slightly different in RAV-6 and tsLA 337 and some recombinants had RAV-6- and others tsLA 337/335-specific CAP oligonucleotides (Table 1, Fig. 2). These oligonucleotides are probably not translated into protein, because they lack an initiation triplet (26) and therefore are not expected to represent viral genes.

The Oligonucleotide Maps of tsLA 3342 and of Four tsLA 3342 X RAV-6 Recombinants. In order to identify directly oligonucleotides segregating with the gag gene, recombinants between tsLA 3342, a mutant of B77 RSV C with a temperature-sensitive gag phenotype (17), and RAV-6 were investigated. The isolation of such recombinants proved to be difficult because most foci selected for infection by virus with the env gene of RAV-6 and the ts gag-marker of LA 3342 appeared to produce heterozygotes consisting of both parental viruses. The viral RNA contained both nd RSV-specific class a RNA and leukosis virus-specific class b RNA (27), and its fingerprint pattern was a superposition rather than a selection of parental oligonucleotides as is typical of recombinants (4,5). However, recently we

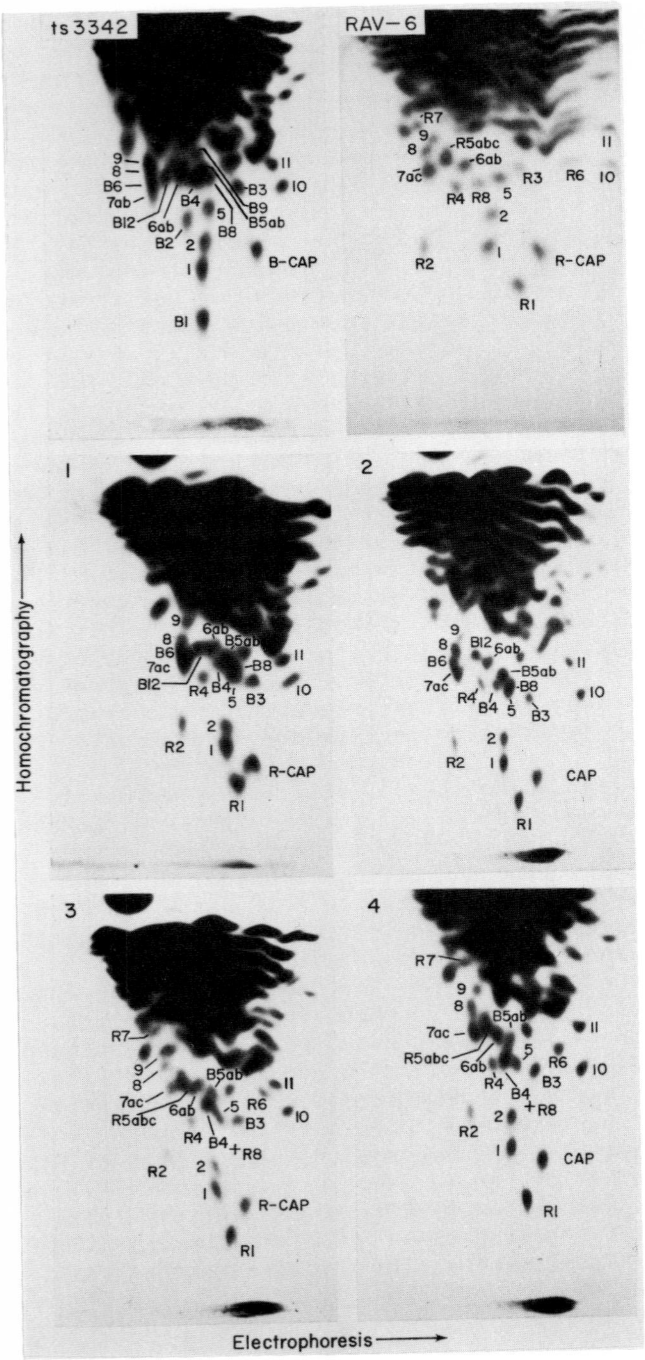

Fig. 3. (See next page for legend)

Fig. 3. RNase T_1-resistant oligonucleotides of the 60-70S (^{32}P) RNAs of B77 RSV C tsLA 3342, leukosis virus RAV-6 and for tsLA 3342 X RAV-6 recombinants (#1-4) after electrophoresis and homochromatography (fingerprinting). Conditions were as described for Fig. 1 except that a yeast RNA solution of 300 A260 and 150 A280 was used instead of the higher concentration of 600 A260 and 300 A280 used above and previously (Fig. 1, ref. 9). Autoradiography was with Kodak x-omatic film using a Dupont "lightening plus" screen. The fingerprint of RAV-6 RNA is the same as that shown in Fig. 1. The B-, R-, and unlettered numbers identify tsLA 3342-, RAV-6-, and common oligonucleotides (Table 1).

succeeded in obtaining recombinants by cloning singly infected and transformed cells in agar suspension (27) to select recombinant virus stocks. Viruses were selected for src of tsLA 3342, the env gene of RAV-6 of subgroup B and either a ts- or wild-type gag. Viruses were judged pure recombinants if their 30-40S RNA species consisted only of size class a (data not shown), as is typical of pure recombinant, nondefective RSV RNA (4,5,27). Four tsLA 3342 X RAV-6 recombinants were analyzed; #1 and #2 had a ts-gag of LA 3342, and #3 and #4 had a wild-type gag of RAV-6 (see below, Fig. 4). Oligonucleotides of tsLA 3342, RAV-6 and the four recombinants are shown in Fig. 3. B77 RSV tsLA 3342-specific oligonucleotides were designated as B-numbers, RAV-6 oligonucleotides as R-numbers, and oligonucleotides shared by both viruses were given unlettered numbers (Table 1, Fig. 3). Analyses of the RNase A-resistant fragments of the large oligonucleotides of each virus are reported in Table 1. These and other data described below (Fig. 4), underscore again the great similarity between B77 (tsLA 3342) and PR C (ts 337) noted previously (9,15).

Oligonucleotide maps of tsLA 3342, of RAV-6, and of four tsLA 3342 X RAV-6 recombinants are shown in Fig. 4. The map of tsLA 3342 differed from that of a wild-type strain of B77 published previously (9), in that it contained a specific oligonucleotide termed here #10, and lacked oligonucleotide #23 of B77. Oligonucleotide maps of one of the recombinants (#2) were derived by determining from discrete sizes of poly(A)-tagged RNA fragments, the distance between each oligonucleotide and the 3' poly(A) coordinate of the viral RNA. The oligonucleotide maps of the remaining recombinants (#1,3,4) were drawn by assigning for each recombinant oligonucleotide a map location which is equivalent to its parental counterpart (see above). The presence or absence of parental oligonucleotides in recombinant oligonucleotide maps as well as the parental origin of distinct recombinant oligonucleotides are depicted in Fig. 4 in the same way as described above for Fig. 2.

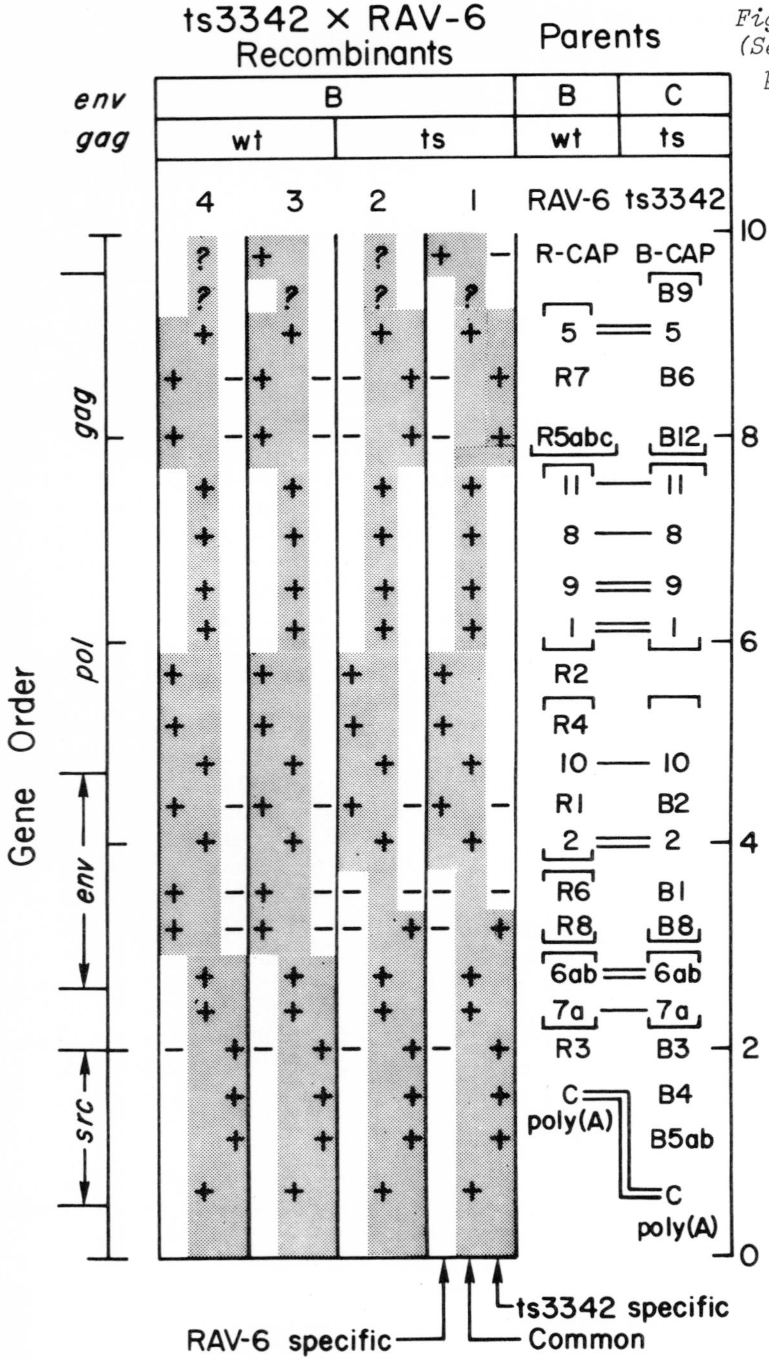

Fig. 4.
(See next page)

Fig. 4. Oligonucleotide maps of tsLA 3342, RAV-6, and of four (#1-4) tsLA 3342 X RAV-6 recombinants. The upper abscissa indicates the env- and gag-phenotypes of each virus. All recombinants also were selected for the src gene of tsLA 3342 (see text). Oligonucleotides are numbered as described for Fig. 3 and Table 1, and have been ordered on the basis of their distance from the 3' poly(A) end of the viral RNA (9). All other symbols are as described for Fig. 2.

Comparing oligonucleotide map segments with the known genetic functions of the parents and their recombinants, we make the following deductions: (i) All four recombinants shared between poly(A) and nucleotide 2000 four oligonucleotides (B3, B4, B5ab) with tsLA 3342. Since B4 and B5ab are src-specific oligonucleotides of B77 (9), and define the only map segment shared by all four recombinants exclusively with tsLA 3342, this map segment must contain the src gene. In addition, all recombinants and both of their parents shared oligonucleotide C, which maps between src and poly(A) (2,9,12). (ii) All four recombinant RNAs contained RAV-6-specific oligonucleotides between 3000 and 6000 nucleotides from the poly(A) end. Since the only known RAV-6 marker shared by all four of these recombinants is the env of subgroup B, it follows that some of these oligonucleotides must be env-specific. This deduction is consistent with the previous location of env in this map segment (2,9,11,12). (iii) All four recombinant RNAs and their parents shared four oligonucleotides located between 6000 and 8000 nucleotides from the poly(A) end. Evidence described above suggested that some or all of these are pol-specific. (iv) The 5'-terminal map segment, between 8000 and 10,000 nucleotides, but not including the 5'-terminal CAP-oligonucleotide, contained two tsLA 3342-specific oligonucleotides (B6 and B12) in recombinants #1 and #2, and contained RAV-6-specific oligonucleotides (R7, R5abc) in recombinants #3 and #4. Since the oligonucleotides of the 5' map segments of the four recombinants are the only ones that segregated with their parental gag markers (e.g., tsLA 3342 gag in recombinants 1 and 2, RAV-6 gag in recombinants 3 and 4), we conclude that these oligonucleotides are probably gag-specific. However more recombinants must be analyzed for a definitive identification of gag-oligonucleotides.

DISCUSSION

Our results provide direct evidence that the gag gene maps between the pol gene and the 5' end of the viral RNA, and demonstrate that the four known genes of RSV have the order 5'-gag-pol-env-src-poly(A). This map is consistent with biochemical and genetic evidence described previously (2,4,7,9-15).

Our identification of the gag gene as a map segment of about 2000 nucleotides near the 5' end of the RNA, which includes strain-specific as well as conserved oligonucleotides, is consistent with the known characteristics of its protein product. The gag gene proteins contain serologically and biochemically conserved as well as strain-specific elements (28,29). The primary gag gene product is a precursor protein of 76,000 daltons (16). This protein is also the only one that is effectively translated in vitro from viral RNA (20). This is an independent argument for the 5' location of the gag gene, since eukaryotic mRNAs are thought to have only a single active initiation site for translation near their 5' end (30).

There is still some uncertainty about the identification of the pol gene, because a viral strain-specific pol oligonucleotide has not yet been identified. Thus the proposed association of pol with a map segment identified by four conserved oligonucleotides between 6000 and 8000 nucleotides from the poly(A) end is indirect, based on the exclusion of other map segments which did not segregate with a ts pol marker. The association of the pol gene with four conserved or highly conserved oligonucleotides is consistent with the notion that the polymerases of different viral strains are serologically (31, 32) and biochemically (23,24) closely related. It may be argued that a map segment between env and src and characterized by oligonucleotides 6ab and 7a, which is also conserved in all virus strains tested here, may be pol-specific. However, this has been considered unlikely (see above), because it appears too small to code for the viral polymerase.

Mechanism of Recombination. RNA tumor viruses are thought to recombine by crossing-over, presumably at the level of proviral DNA (4,5). Little is known about the detailed mechanism. The oligonucleotide maps of our recombinants suggest a minimum of two to five cross-over points (Figs. 2 & 4). (Cross-overs that do not involve RNase-T_1-resistant oligonucleotides would not be detected by our method.) Thus far, our data do not suggest that RNA segments exist that are especially recombination prone. However, some peculiarities were noted; all 20 tsLA 337/335 X RAV-6 recombinants lacked the oligonucleotide P10 of tsLA 337/335. A linkage between gag and src has been observed in genetic experiments (17,18) reviewed previously (13). Such a linkage is not suggested by biochemical data reported here and previously (12,13,14,15). In addition, no more than 15 out of the 24 recombinants studied here appear to contain sarcoma parental CAP-oligonucleotides (preliminary observation). The apparent discrepancy may reflect a technical bias. Recombinants studied biochemically must be from biochemically distinguishable parents, whereas recombinants studied genetically may be from biochemically indistinguishable, congenic parents differing only in conditional lesions. The linkage patterns of gag and src may be different in recombination between biochemically distinct and congenic parents.

ACKNOWLEDGMENT

This work was supported by Public Health Service research grants CA-19558 and CA-11426 from the National Cancer Institute.

REFERENCES

(1) Baltimore, D. (1975) Cold Spring Harbor Symp. Quant. Biol. 39, 1187.
(2) Wang, L.-H., Duesberg, P.H., Kawai, S. & Hanafusa, H. (1976) Proc. Natl. Acad. Sci. USA 73, 447.
(3) Billeter, M.A., Parsons, J.T. & Coffin, J.M. (1974) Proc. Natl. Acad. Sci. USA 71, 3560.
(4) Beemon, K., Duesberg, P.H. & Vogt, P.K. (1974)Proc. Natl. Acad. Sci. USA 71, 4254.
(5) Duesberg, P.H., Vogt, P.K., Beemon, K. & Lai, M. M.-C. (1975) Cold Spring Harbor Symp. Quant. Biol. 39, 847.
(6) Lai, M.M.-C. & Duesberg, P.H. (1972) Nature 235, 383.
(7) Wang, L.-H. & Duesberg, P.H. (1974) J. Virol. 14, 1515.
(8) Lai, M.M.-C., Duesberg, P.H., Horst, J. & Vogt, P.K. (1973) Proc. Natl. Acad. Sci. USA 70, 2266.
(9) Wang, L.-H., Duesberg, P.H., Beemon, K. & Vogt, P.K. (1975) J. Virol. 16, 1051.
(10) Coffin, J.M. & Billeter, M.A. (1976) J. Mol. Biol. 100, 293.
(11) Joho, R.H., Billeter, M.A. & Weissmann, C. (1975) Proc. Natl. Acad. Sci. USA 72, 4772.
(12) Wang, L.-H., Duesberg, P.H., Mellon, P. & Vogt, P.K. (1976) Proc. Natl. Acad. Sci. USA 73, 1073.
(13) Duesberg, P.H., Wang, L.-H., Mellon, P., Mason, W.S. & Vogt, P.K. (1976) Proceedings of the ICN-UCLA Symposium on Animal Virology, eds. Baltimore, D., Huang, A. & Fox, C.F. (Academic Press, New York), Vol.4, p.107.
(14) Wang, L.-H., Galehouse, D., Mellon, P., Duesberg, P.H., Mason, W.S. & Vogt, P.K. (1976) Proc. Natl. Acad. Sci. USA 73, 3952.
(15) Joho, R.H., Stoll, E., Friis, R.R., Billeter, M.A. & Weissmann, C. (1976) Proceedings of the ICN-UCLA Symposium on Animal Virology, eds. Baltimore, D., Huang, A. & Fox, C.F. (Academic Press, New York), Vol. 4, p.127.
(16) Von der Helm, K. & Duesberg, P.H. (1975) Proc. Natl. Acad. Sci. USA 72, 614.
(17) Hunter, E. & Vogt, P.K. (1976) Virology 69, 23.

(18) Hayman, M.J. & Vogt, P.K. (1976) Virology 73, 372.
(19) Wyke, J., Bell, J.G. & Beamand, J.H. (1975) Cold Spring Harbor Symp. Quant. Biol. 39, 897.
(20) Eisenman, R., Vogt, V.M. & Diggelmann, H. (1975) Cold Spring Harbor Symp. Quant. Biol. 39, 1067.
(21) Mason, W.S., Friis, R.R., Linial, M. & Vogt, P.K. (1974) Virology 61, 559.
(22) Vogt, P.K. (1971) Virology 46, 947.
(23) Moelling, K. (1975) Cold Spring Harbor Symp. Quant. Biol. 39, 969.
(24) Gibson, W. & Verma, I.M.(1974) Proc. Natl. Acad. Sci. USA 71, 4991.
(25) Keith, J. & Fraenkel-Conrat, H. (1975) Proc. Natl. Acad. Sci. USA 72, 3347.
(26) Beemon, K. & Keith, J. (1976) Proceedings of the ICN-UCLA Symposium on Animal Virology, eds. Baltimore, D., Huang, A. & Fox, C.F. (Academic Press, New York) Vol. 4, p.97.
(27) Duesberg, P.H. & Vogt, P.K. (1973) Virology 54, 207.
(28) Huebner, R.J., Armstrong, D., Okuyan, M., Sarma, P. & Turner, H. (1964) Proc. Natl. Acad. Sci. USA 51, 742.
(29) Bolognesi, D.P., Ishizaki, R., Hüper, G., Vanaman, T.C. & Smith, R.E. (1975) Virology 64, 349.
(30) Jacobson, M.F. & Baltimore, D. (1968) Proc. Natl. Acad. Sci. USA 61, 77.
(31) Parks, W.P., Scolnick, E.M., Ross, J., Todaro, G.J. & Aaronson, S.A. (1972) J. Virol. 9, 110.
(32) Nowinski, R.C., Watson, K.F., Yaniv, A. & Spiegelman, S. (1972) J. Virol. 10, 959.
(33) Hunter, E., Hayman, M.J., Rongey, R.W. & Vogt, P.K.(1976) Virology 69, 35.

Discussion

C. Weissman, Universität Zurich: I think one must be careful about relying on only one oligonucleotide change in this type of analysis, because these viruses have a very high rate of mutation. A single nucleotide substitution can lead to the disappearance of an oligonucleotide and thus lead to the erroneous conclusion that a recombination has occurred.

P. Duesberg, University of California, Berkeley: Yes, in the gag-gene analyzed here there are actually two specific oliognucleotide spots - one of them is a double-spot. I agree - we had that problem when we first tried to correlate a single oligonucleotide with a product of the gag-gene the P27 and I think there we run into difficulties, slicing the salami too thin for this method. We look only at oligonucleotides which altogether are only 5% of the RNA and each one of those gag proteins takes up only about 5-10% of coding capacity of the RNA. Consequently, we look at proteins which may not even have an oligonucleotide counterpart as determined by our method. So I agree with you. We should have at least two coordinately segregating oligonucleotides to define a genetic marker.

S. O'Brien, National Institutes of Health: When you select for recombinants between the envelope and the SVC gene, what percent of the recombinants recovered actually involve more than one exchange?

P. Duesberg: That is a difficult question to answer because by the time we look at them they have been cloned and have undergone multiple cycles of replication. The numbers that we have so far range between one and five. I cannot say whether this reflects the original distribution of the primary recombination event and how much secondary recombinations have contributed.

RECOMBINATION EVENTS BETWEEN SIMIAN VIRUS 40 AND THE HOST GENOME

E. WINOCOUR, M. OREN, S. LAVI, T. VOGEL and Y. GLUZMAN
Department of Virology, Weizmann Institute of Science
Rehovot, Israel

Recombination between the cellular genome and that of the tumor virus may play an important role not only in the integration events that lead to neoplastic transformation, but also in the way new types of cancer viruses evolve in nature. In this report, we will discuss 2 aspects of host-virus recombination in the Simian virus 40 (SV40) system. The first is concerned with the structure of a host-substituted SV40 variant, in which large portions of the viral genome have been replaced by covalently-linked cellular DNA sequences. The second aspect deals with recombination between exogenous and endogenous SV40 genes.

A. THE STRUCTURE OF A CLONED SUBSTITUTED SV40 VARIANT.

During the replication of SV40 in monkey cells, cellular DNA is covalently incorporated into defective viral genomes known as substituted variants (reviewed in Ref. 1). The relationship between the mechanism which generates host-substituted SV40 DNA and the integration events leading to neoplastic transformation is unclear. Nevertheless, it has been of interest to determine the types of host sequences in different substituted SV40 populations since the variety of such sequences should reflect the number and nature of the host-virus recombination sites. We have reported that a particular group of host sequences, derived predominantly from the non-reiterated fraction of the cellular genome, is acquired with a strikingly high frequency by different independently-generated populations of substituted SV40 (2). The history of these populations is described in Fig. 1. Three of the serially-passaged defective SV40 populations (denoted as CVB-1-P4, CVB-4-P8, and CVG-1-P10 in Fig. 1) were found to share a common set of host sequences. Furthermore, a part of this common set was also represented in 3 other independently-derived serially-passaged populations of SV40 strain 777 and in SV40 strains 776 and Rh911. From these observations we concluded that the host-virus recombination process occurs at preferred sites on the monkey genome. The objective of the experiments described below was to clarify the structure of the substituted SV40 genome that contains the common, prevalent host sequences.

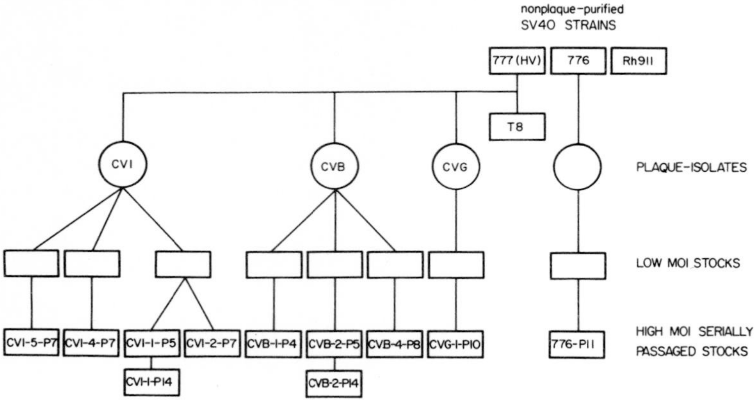

Fig. 1. The passage history of the substituted SV40 stocks. Reprinted, with permission, from Oren, Kuff and Winocour (2).

The cloned substituted SV40 variant was isolated by the procedure of Brockman and Nathans (3) from the CVB-1-P4 population whose history is described in Fig. 1. The mixed population, containing standard SV40 and defective variants, was plated at 10^{-6} dilution and 24 well-separated plaques were picked. Twenty-three of the 24 plaques contained only standard-sized SV40 DNA, as judged by gel electrophoresis. One plaque contained, in addition to supercoiled standard SV40 DNA, a single species of supercoiled defective DNA, 85% the size of standard DNA. The smaller supercoiled defective DNA (designated F161-F) lacked the E. coli (Eco) R1 cleavage site and hybridized to both viral and host DNA. A radioactive cRNA probe (2) directed against the host sequences in F161-F showed a pattern of hybridization to other substituted SV40 stocks similar to that of the parental uncloned CVB-1-P4 population (Table 1). The specific group of host sequences which is found with high frequency in different independently-generated substituted SV40 stocks is thus also represented in the cloned F161-F variant DNA.

TABLE 1

Common host sequences in cloned (F161-F) and various uncloned populations of substituted SV40 DNA*

DNA on filter 1 μg	^{3}H-cRNA ("host-sequence probe")	
	CVB-1-P4	F161-F
SV40:		
Plaque-purified 777	2.2	2.4
CVB-1-P4	83.1	97.7
CVG-1-P10	87.5	93.0
CV1-1-P5	3.7	3.4
Rh 911	17.2	17.0
Monkey BSC1 cells	15.0	16.7

*The passage history of the different SV40 stocks is shown in Fig. 1. F161-F is a clonal population of substituted SV40 DNA molecules derived from the CVB-1-4 stock (see text). The figures in the table refer to the percentage of the ^{3}H-cRNA input (7000 cpm) which hybridizes with the indicated DNA on filters. See (2) for details of the preparation of the ^{3}H-cRNA "host-sequence probe" (directed against the host sequences in CVB-1-P4 and F161-F) and the hybridization procedure used.

The structure of the F161-F genome was investigated by restriction endonuclease mapping and hybridization techniques. A comparison of the Haemophilus influenzae (HinII+III) restriction endonuclease cleavage maps of standard SV40 DNA and F161-F DNA is shown in Fig. 2. Cleavage of standard plaque-purified SV40 DNA by HinII+III generates 11 main classes of fragments in yields which are equimolar with the original DNA (4). Cleavage of F161-F DNA by HinII+III generates only 7 classes of fragments. The data in Table 2 shows that on the basis of size and percentage yield, only 3 of the 7 classes are equimolar with the original DNA; the molarity of the remaining classes is greater than one, suggesting that these fragments are represented more than once per DNA molecule. Indeed, the data in Table 2 indicate that fragments 1, 3 and 4 are repeated 3 times, and fragment 6, 4 times in the F161-F genome.

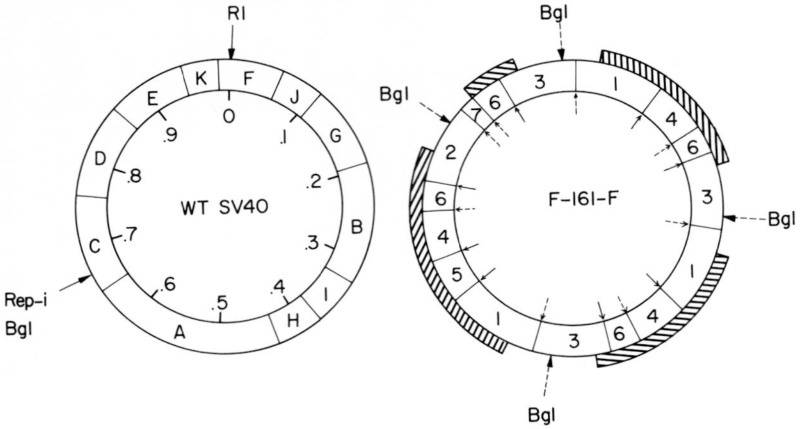

Fig. 2. HinII+III cleavage maps of the wild-type standard SV40 genome (left) and the cloned substituted F161-F genome (right). In the map of wild-type SV40, R1 and Rep-i refer, respectively, to the sites of Eco R1 cleavage and the initiation of replication (4); Bgl refer to the cleavage site of endonuclease BglI (S. Zain, personal communication). The map of F161-F is based partly upon the data shown in Table 2 and partly upon studies to be reported elsewhere (5). The order of fragments 1-7 was determined by sequential cleavage (4) with HinII followed by HinIII and *vice versa*. The inner arrows denote the HinII (→) and HinIII (--→) cleavage sites. The shaded segments denote the presence of host sequences (see text).

TABLE 2

The HinII+III cleavage products of F161-F*

Fragment	Percent unit-length	Percent yield	Molarity	Percent viral sequences
1	8.5	30.4	3	37
2	7.6	7.7	1	74
3	7.3	21.6	3	76
4	4.8	15.3	3	0
5	4.0	4.4	1	0
6	2.8	13.1	4	0
7	1.4	1.5	1	NT

*The fragments were separated by electrophoresis on 5% polyacrylamide gels. The cleavage products of the standard SV40 genome were run in the same gels as markers. Percent unit-length was calculated from the electrophoretic mobility relative to that of the markers. Percent yield refers to the percent of the radioactivity recovered in each band. Molarity = percent yield/percent unit-length multiplied by 0.85 (the F164-F genome is 85% the length of the standard SV40 genome). Percent viral sequences was determined in S1-monitored hybridization reactions between fragments 1-6 and excess plaque-purified standard SV40 DNA as described in (6). Data compiled from Oren, Lavi and Winocour (5).

Evidence for the presence of repeated segments in the F161-F genome has also been obtained by measuring the reassociation (self-annealing) rates of single-stranded DNA. The initial portion of the Cot curve indicates that some sequences in F161-F DNA reassociate as much as 15-20 times faster than standard SV40 DNA; at the Cot½ level, the average reassociation rate is 4-5 times faster than SV40 DNA; and it is only at the higher Cot values (>70% reassociated) that the reassociation curves of SV40 and F161-F DNAs begin to converge, indicating that only a small proportion of sequences in F161-F are not repeated. Reassociation rate studies with the individual classes of fragments generated by HinII+III cleavage of F161-F DNA have also indicated the presence of repeated sequences <u>within</u> the fragments which contain predominantly viral sequences (fragments 2 and 3; see below). The significance of this is currently being investigated. The presence of repeated viral sequences in the parental uncloned populations from which F161-F was isolated, has been reported previously (7).

The proportion of viral sequences in the HinII+III fragments of F161-F DNA was determined in hybridization reactions with plaque-purified standard SV40 DNA (final column of Table 2). Fragments 4, 5, and 6 contained no detectable levels of viral sequences and we assume that these fragments contain exclusively host sequences (see below). To determine which part of the standard SV40 genome is represented in fragments 1, 2, and 3, we measured the ability of labelled F161-F DNA to hybridize with different segments of the standard SV40 genome, derived by cleaving SV40 DNA with various restriction endonuclease combinations.

TABLE 3

The portion of the standard SV40 genome represented in F161-F DNA*

SV40 cleaved by	Fragment		
	Fractional length	Hybridization	
		(^{32}P) SV40	(^{32}P) F161-F
BamI+HhaI +R1	0-0.14	+	-
	0.14-0.74	+	+
	0.74-0.84	+	-
	0.84-0	+	-
HaeIII	0.28-0.59	+	-
HinIII + HpaII	0.64-0.72	+	+

*Standard plaque-purified SV40 DNA was digested with the indicated combination of restriction endonucleases. The cleavage products were separated on 1.4% agarose slab gels and transferred to cellulose nitrate filters by Southern's "blotting" technique (8). DNA-DNA hybridization with (^{32}P) F161-F DNA or plaque-purified standard (^{32}P) SV40 DNA was monitored by radioautography. The fractional lengths of the indicated fragments refer to the distance from the Eco R1 site (4). Data compiled from (5).

Table 3 shows the fractional lengths (distance from the Eco Rl site) of the SV40 genome segments and the presence or absence of the corresponding sequences in F161-F DNA (the control in each case being provided by the positive hybridization with the homologous ^{32}P-labelled standard SV40 DNA probe). The combination of enzymes BamI (Bacillus amyloliquefaciens) + HhaI (Haemophilus haemolyticus) + Rl splits the SV40 genome into 4 segments (4); and the data show that the viral sequences in F161-F DNA are only contained within one of these segments (0.14-0.74 map units). However, only a relatively small part of the 0.14-0.74 segment contains the F161-F sequences since the Haemophilus aegyptius (Hae)III fragment A (0.28-0.59 units) does not hybridize with F161-F DNA. On the other hand, the segment generated by HinIII+ HpaII (Haemophilus parainfluenzae), 0.64-0.72 units, hybridizes with F161-F DNA. From these results, we conclude that the viral sequences in the F161-F genome are exclusively derived from within the segment 0.59-0.74 units from the Eco Rl site in the standard SV40 genome. Since this segment contains the origin of replication (at 0.67 units), we can also conclude that the F161-F genome contains multiple origins of replication. The presence of multiple origins of replication in defective SV40 genomes, including substituted variants, has been reported previously (1, 7,9-13).

The 0.59-0.74 segment of the standard SV40 genome contains the sites of the BglI (Bacillus globiggi), HhaI and HpaII restriction endonucleases (S. Zain, personal communication and Ref. 4). It was thus of interest to map the cleavage sites of these enzymes in the F161-F genome. BglI cleaves F161-F DNA into 3 size-classes of fragments (of molarities 1, 2 and 1) and the positions of the 4 cleavage sites are shown in Fig. 2. The cleavage sites of HpaII and HhaI are also located in the viral portions of fragments 3 and 2 (data not shown); and the molarities of the cleavage-product yields are consistent with the repeat-structure shown in Fig. 2. Thus, the mapped cleavage-site positions of the BglI, HhaI and HpaII enzymes on the F161-F genome provide confirmation of the structure shown in Fig. 2.

The final column of Table 2 shows that fragments 4, 5, 6 and parts of fragments 1, 2, and 3 contain sequences that fail to hybridize with the standard SV40 genome. Complementary RNA transcribed from F161-F DNA hybridizes with reiterated monkey cell DNA (Table 1). The experiment described schematically in Fig. 3 provides evidence that the reiterated host sequences are located mainly in fragment 6.

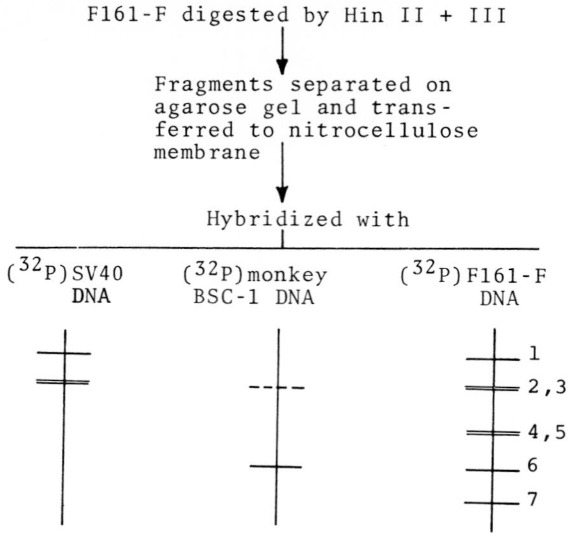

Fig. 3. Position of the reiterated monkey sequences in F161-F DNA. F161-F DNA was digested with HinII+III. The 7 classes of fragments were separated on 1.4% agarose slab gels and hybridized with (^{32}P) monkey BSC-1 cell DNA, (^{32}P) SV40 DNA, and (^{32}P) F161-F DNA, using Southern's "blotting" technique as described under Table 3. The drawing indicates the fragment classes which hybridized with each probe, as determined by radioautography of the "gel blots". The hybridization of (^{32}P) monkey BSC-1 DNA to fragments 2+3 (which are poorly separated on the gel) is uncertain and is denoted by a broken line. Compiled from (5).

---

^{32}P-labelled monkey BSC-1 cell DNA was hybridized with the 7 classes of fragments generated by the digestion of F161-F with HinII+III. Under the conditions used, only reiterated host DNA sequences are expected to hybridize. The ^{32}P-monkey DNA hybridized clearly with fragment 6 (and possibly also with parts of fragments 2 and 3). The results of the experiment described in Fig. 3 also confirm the data in Table 2 in that the ^{32}P-SV40 DNA probe hybridized only with fragments 1, 2 and 3. We are now trying to establish by direct hybridization with excess monkey DNA that our assumption of unique host sequences in fragments 1, 4 and 5 is correct.

The information obtained so far on the structure of the F161-F genome may be summarized as follows. The viral sequences (~39% of the total) are distributed in 4 noncontiguous segments, 3 of which appear to be repeat units of approximately 450 base pairs. The 4th segment, represented by parts of fragments 2 and 7 (Fig. 2) shares common sequences with the other viral segments, but is smaller. Each of the 4 segments contains one BglI, one HpaII, and one HhaI cleavage site. Only a limited part of the standard SV40 genome is represented in F161-F DNA; this part being contained within the maximum limits, 0.59-0.74 units from the Eco R1 site. Possibly, the part of the standard genome represented is even smaller and contains repeated sequences. Assuming that the remaining parts of the F161-F genome which hybridize neither to viral DNA nor to reiterated host DNA contain unique host sequences, 61% of the total sequences in F161-F are of host origin, and are interspersed between the 4 viral repeat units. Preliminary data indicate that the 4 host DNA segments share common sequences. We estimate that each of the 3 larger segments contains approximately 200 base pairs of reiterated host DNA and about 500 base pairs of unique host DNA. The potential informational content of the host DNA in the F161-F genome is thus limited and sufficient to code for only about 16,000-18,000 daltons of protein.

The type of substituted SV40 genome represented by the cloned F161-F variant is frequently found in different defective SV40 stocks. A similar structure of interspersed viral and host repeat units has also been observed by Lee, Brockman and Nathans (13) and by M.F. Singer (personal communication). If such genomes are derived initially from integrated SV40 (1), then the generation of the F161-F and similar genomes must involve 2 primary steps; the excision of the basic subunit (monomer) and the amplification of the subunits (13). Our results with F161-F have indicated, however, that the "repeat symmetry" is imperfect in that fragments 5, 2 and 7 occur only once per DNA molecule; and only 3 of the 4 proposed viral repeat units and 2 of the 4 proposed host DNA repeat units in Fig. 2 are of equal size.

B. RECOMBINATION BETWEEN ENDOGENOUS AND EXOGENOUS SV40 GENES.

Certain properties of the RNA tumor viruses change after passage in particular host cells (reviewed in Ref. 14). Since most, if not all, eukaryotic genomes harbor endogenous RNA virus sequences, recombination between infecting viruses and endogenous integrated viral genes may account for the observed heterogeneity in the properties of these viruses

(14, 15). Indeed, exogenous-endogenous viral recombination events may well be responsible for the evolution of new cancer viruses with altered oncogenic potential. To study this type of recombination at the molecular level, we have established the following SV40 system.

The system is based upon the rescue of SV40 temperature-sensitive (ts) mutants after passage at the permissive temperature, on permissive transformed monkey cell lines whose resident SV40 genome is wild-type with respect to the super-infecting ts-mutant. Marker-rescued virus is isolated from plaques produced by infecting normal monkey cells at the restrictive temperature with the passaged ts-virus. The DNA of the marker-rescued virus is then analyzed by restriction endonuclease mapping to provide evidence that it contains segments characteristic of both the endogenous SV40 genome of the transformed cells and the exogenous superinfecting ts-mutant.

We shall first summarise the properties of the permissive SV40-transformed monkey cells (16) which are crucial to the interpretation of the marker-rescue experiments. The monkey cells, of the permissive CV1 line, were transformed by UV-inactivated SV40 of strain 777. This strain appears to be a late ts-mutant since it plaques poorly at 40.5^{o}C and can be complemented by the early tsA30 mutant, but not by the late tsB204 mutant (17). Three lines of transformed cells which were initially selected in soft agar containing anti-SV40 antiserum, were found to be fully sensitive to superinfection. The transformed lines themselves released no infectious virus. By Cot analysis, the permissive transformed monkey cells were found to contain, on the average, 1-2 SV40 genome equivalents associated with high molecular-weight chromosomal DNA. Since the transformed monkey cells failed to support the replication of SV40 tsA mutants at the restrictive temperature (they fully support the growth of tsA mutants at the permissive temperature), we have suggested that the integrated genome, although competent with respect to transformation, is defective with respect to the initiation of viral DNA synthesis (16).

Experiments with the SV40 uncoating mutant tsD202 have shown that passage of the mutant virus at the permissive temperature on each of the 3 permissive lines of SV40 transformed monkey cells gives rise to progeny whose efficiency of plating at the restrictive temperature on normal monkey CV1 cells is comparable to that of wild-type SV40 (17). Compared to the control tsD202 population passaged twice on normal monkey cells, the plating ratio (number of plaques at 40.5°C/number of plaques at 33.5°C) of the same virus passaged twice on transformed cells increased by 10^3 to 10^6-fold. The rescued tsD202 virus retains its high efficiency of plating at 40.5°C following sequential plaque-purification and displays a typical single-hit dosage-response curve at the restrictive temperature. At the permissive temperature (33.5°C) it possesses no selective growth advantage on transformed cells compared to normal monkey cells. Hence, the repair of the ts-phenotype is neither due to complementation effects nor to the selection of pre-existing tsD202 revertants which grow better on the transformed cells.

The following lines of evidence show that the rescue of the tsD202 phenotype is due to recombination with the endogenous SV40 genome of the transformed cells (18).

I. The ts-lesion in D202 is located within Hin fragment E (Fig. 2, left panel) at ~0.9 map-units from the Eco Rl site (19, 20). Hin fragment E of wild-type SV40 contains a HaeIII cleavage site, located at approximately 0.9 map-units (see Ref. 4). We have shown that tsD202 DNA lacks the HaeIII cleavage site contained within Hin fragment E. In contrast, the DNA of rescued D202, like that of the SV40 777 used to transform the monkey cells, contains this HaeIII cleavage site.

II. Comparison of the HinII+III cleavage patterns of SV40 777 and tsD202 DNAs by gel electrophoresis have indicated differences in the mobilities of fragments A, B and F. The electrophoretic mobilities of fragments A and B from SV40 777 DNA are lower than those of the same fragments from tsD202 DNA; the mobility of fragment F from SV40 777 DNA is higher than that of fragment F from tsD202 DNA. The HinII+III cleavage pattern of rescued D202 DNA shows that the mobility of the A, B and F fragments is identical to that of those from tsD202 DNA.

Observation I implies that the D202 genome rescued by passage on SV40-transformed monkey cells acquired a segment from the endogenous SV40 777 DNA which contains the HaeIII cleavage site located at approximately 0.9 map-units. It will be recalled that the lesion of the ts-parental virus maps at the same position (19,20). Observation II indicates that other regions of the rescued D202 genome (defined by Hin fragments A, B and F) are characteristic of the parental tsD202 DNA. Furthermore, since the SV40 777 used to transform the monkey cells is a late ts-mutant, the segment of rescued D202 defined by Hin fragments K-G (where the ts-lesions of SV40 late mutants map; 19, 20) cannot be entirely of 777 origin. Thus, the rescued genome contains segments from both parents - the exogenous tsD202 virus and the endogenous SV40 genome of the transformed cells (Table 4). Four out of 5 rescued D202 genomes, each derived from independent passages in the 3 SV40-transformed monkey lines, displayed the hybrid characteristics described above. The genome of the 5th isolate lacked the HaeIII cleavage site; the nature of this virus is presently being investigated. In parallel control experiments, the DNA of 3 independently-isolated tsD202 revertants, which appeared at very low frequency when tsD202 virus was passaged on normal monkey CV1 cells, was analysed by the same techniques. The revertant DNAs could not be distinguished from the parental tsD202 DNA by any of the markers described above.

TABLE 4

Comparison of the endogenous, exogenous, and recombinant SV40 viruses

SV40 777 (endogenous virus)	1. The virus used, after UV-inactivation, to transform monkey CV1 cells.
	2. Does not plaque at 40.5°C (complemented by tsA but not by tsB mutants).
	3. Contains the HaeIII cleavage site, within Hin fragment E, at ~0.9 map-units.
SV40 tsD202 (exogenous virus)	1. Uncoating mutant which does not plaque at 40.5°C.
	2. ts-lesion maps at ~0.9 map-units.
	3. Lacks the HaeIII cleavage site at ~0.9 map-units.
	4. Hin fragments A and B migrate faster than those of SV40 777.
	5. Hin fragment F migrates slower than that of SV40 777.
SV40 D202 recombinant	1. Accumulates during passage of tsD202 at permissive temperature on SV40-transformed CV1 lines.
	2. Plaques at 40.5°C.
	3. Contains the HaeIII cleavage site at ~0.9 map-units.
	4. Hin fragments A, B and F migrate like those of tsD202 DNA.

Marker-rescue experiments, similar to those with the tsD mutant, have been performed with the tsA, tsB and tsC classes of mutants. No significant level of rescue was observed with the tsB and tsC viruses. The experiments with the tsA mutants showed a degree of rescue that was only slightly higher than the spontaneous reversion rate. The lack of detectable rescue with the tsB and tsC mutants can be explained on the basis that the resident SV40 genome in the transformed cells is itself a late mutant. The very low level of rescue obtained with the tsA class of mutants is consistent with our suggestion that the endogenous SV40 genome is defective in the function required for the initiation of viral DNA synthesis (16).

REFERENCES

(1) E. Winocour, N. Frenkel, S. Lavi, M. Osenholts (Oren) and S. Rozenblatt. Cold Spring Harbor Symp. Quant. Biol. 39 (1975) 101.

(2) M. Oren, E.L. Kuff and E. Winocour. Virology 73 (1976) 419.

(3) W.W.Brockman and D. Nathans. Proc. Nat. Acad. Sci. USA 71 (1974) 942.

(4) D. Nathans and H.O. Smith. Ann. Rev. Biochem. 44 (1975) 273.

(5) M. Oren, S. Lavi and E. Winocour. Submitted for publication.

(6) N. Frenkel, S. Lavi and E. Winocour. Virology 60 (1974) 9.

(7) N. Frenkel, S. Rozenblatt and E. Winocour, in: Tumor Virus-Host Cell Interaction, ed. A. Kolber (Plenum Publishing Corporation, New York, 1975) p. 39.

(8) E.M. Southern. J. Mol. Biol. 98 (1975) 503.

(9) W.W. Brockman, T.N.H. Lee and D. Nathans. Cold Spring Harbor Symp. Quant. Biol. 39 (1975) 119.

(10) D. Davoli and G.C. Fareed. Ibid 39 (1975) 137.

(11) M.A. Martin, G. Khoury and G.C. Fareed. Ibid 39 (1975) 129.

(12) J.E. Mertz, J. Carbon, M. Hertzberg, R.W. Davis and P. Berg. Ibid 39 (1975) 69.

(13) T.N.H. Lee, W.W. Brockman and D. Nathans. Virology 66, (1975) 53.

(14) H.M. Temin. Adv. Cancer Res. 19 (1974) 47.

(15) R.A. Weiss, W.S. Mason and P.K. Vogt. Virology 52 (1973) 535.

(16) Y. Gluzman, J. Davison, M. Oren and E. Winocour. Submitted for publication

(17) Y. Gluzman, E.L. Kuff and E. Winocour. Submitted for publication.

(18) T. Vogel, Y. Gluzman and E. Winocour. Submitted for publication.

(19) C-J. Lai and D. Nathans. Virology 66 (1975) 70.

(20) T.E. Shenk, C. Rhodes, P.W.J. Rigby and P. Berg. Cold Spring Harbor Symp. Quant. Biol. 39 (1975) 61.

ACKNOWLEGEMENT

The research cited herein was supported by grants from the United States-Israel Binational Science Foundation (BSF), the Israel Ministry of Health, the German Science Fund (GSF) and by a contract (NOI CP 33220) from the National Cancer Institute.

Discussion

P. Duesberg, University of California, Berkeley: Do you know whether these sequences are transcribed?

E. Winocour, The Weizmann Institute of Science: We have not been able to detect transcription of the host DNA sequences in substituted SV40 DNA by the conventional procedure of isolating virus-specific RNA species from intact cells. If, however, the viral RNA is first isolated in the form of a "transcription complex" (either by the Sarkocyl or Triton-X procedure) and RNA synthesis is allowed to proceed in vitro, then the transcription of the host DNA sequences can be readily detected. It is possible, therefore, that in the intact cell, the transcript of the substituted genome is rapidly degraded.

J. Hassell, Cold Spring Harbor: How much of the genetic information in the recombinant D202 is composed of 777 DNA sequences?

E. Winocour: We have not mapped the recombinant genome to the extent that I can tell you precisely which parts have been exchanged. So far, the experiments show only that the recombinant virus contains some sequences derived from the resident SV40 777 genome.

J. Hassell: I have another question. What fraction of the plaques picked at 40° are recombinants? Are they all recombinants or are some revertants?

E. Winocour: We have analyzed 3 revertants from passages on normal cells and 5 recombinants from passages on transformed cells. The 3 revertants were all identical with respect to the markers that I described. 4 out of the 5 independently isolated recombinants (from passages in 3 different lines of transformed monkey cells) were similar in that (a) they had acquired the Hae III site characterisitic of the resident 777 genome and (b) their Hin A, B, and F fragments were characteristic of the ts-parent. The 5th isolate had not acquired the Hae III site; but we do not yet know if this is a revertant or another type of recombinant. So, at least 4 out of 5 plaques picked at 40°C contained recombinant DNA.

J. Schultz, Papanicolaou Cancer Research Institute: Dr. Winocour, for those of us who cannot keep up with the new language now being developed, I would like to ask a simplified question. In this paper, when you transform a kidney cell and then rescue the virus, does the sequence of the rescued virus contain some of the DNA of the monkey host cell? I assume by "rescue" that you mean recovery of the integrated virus. When you pass a virus through a transformed cell which apparently has in its DNA some of the previous virus that caused the cell to be transformed, does the rescued virus contain host DNA sequences?

E. Winocour: I described two different recombinant phenomena which may or may not be related. One is the acquisition of host DNA during passage of SV40 on monkey cells in a lytic infection. Whether that has anything to do with transformation is still an open question. We do not know if the recombination events that generate the host substitution during lytic infection are the same as the recombination events involved in integration and neoplastic transformation. The second type of recombination was between the resident SV40 genome of the transformed monkey cell and a superinfecting SV40 ts-mutant. As a third possible type of recombination event, I think it is certainly feasible that excision of integrated SV40 DNA from a tumor or transformed cell could give rise to host DNA sequences in the rescued virus. But we have not demonstrated this.

M. Singer, National Institutes of Health: We have been working on host substituted defective SV40s that are derived from those isolated in Ernest Winocour's lab. In confirmation of the detection of host repetitive sequences in such substituted viruses, we have found that the sequence of a piece of repetitive monkey DNA, cut out of a defective variant, a piece 150 nucleotide residues long, is identical by sequence determination to 150 residues that can be cleaved directly out of monkey DNA from uninfected and untransformed cells by restriction digestion. In brief, therefore, 2 sequences of 150 residues have been determined separately and shown to be identical: one from one of these substituted defective viruses and one directly from a monkey genome.

THE IMPORTANCE OF TRANSCRIPTION UNIT DEFINITION

J.E. DARNELL, J. WEBER, S. GOLDBERG, W. JELINEK,
N. FRASER and P. SEHGAL
Molecular Cell Biology Dept.
The Rockefeller University
1230 York Ave.
New York, N. Y. 10021

In prokaryotic cells transcription of mRNA is followed by immediate translation of the genetic information into protein. In addition, the mRNA molecule has a short half-life. Thus by controlling transcription alone, a bacterium can control gene expression. In eukaryotic cells there are a number of metabolic events which occur between mRNA transcription in the cell nucleus and the eventual translation of mRNA in the cell cytoplasm. In most eukaryotic mRNA molecules the 5' terminus is modified by a blocked, methylated, oligonucleotide structure termed a "cap" (1) and the 3' terminus is extended by a segment of polyadenylic acid over 200 nucleotides long (2). Both these terminal modifications as well as some sequences present in mRNA molecules can be found in nuclear molecules that are larger than cytoplasmic mRNA (3,4,5). It therefore is possible that primary RNA transcripts are first synthesized, then specifically cleaved to yield new termini, either 5' or 3' or both, before the aforementioned post-transcriptional terminal modifications. Finally mRNA molecules with both short and long half-lives exist, although it is not definitely established whether or not a single mRNA species can have a variable half life (6). Thus, regulation of genetic expression in eukaryotes might, as in bacteria, lie at the level of transcription, or perhaps at some one or more of the post-transcriptional steps in mRNA production.

If an understanding is ever to be gained of which step or steps serve as regulatory sites in mRNA production, specific mRNA sequences must be traced from transcription through to successful cytoplasmic function. The first experimental need to satisfy this long-term goal is, in fact, to define either the boundaries of <u>transcription unit(s)</u> as a region in the DNA which contains specific mRNA sequences or to define the primary RNA product of such transcription units.

The initiation point(s) of transcription for various regions of bacterial or bacteriophage chromosomes were defined first by genetic means. For several operons the genetic predictions have been elegantly confirmed by isolating the relevant DNA: 1) the DNA site which binds RNA polymerase (a "promoter") and in which RNA chain initiation occurs has been isolated and sequenced in the lactose operon (7), in bacteriophage lambda DNA (8) and at the beginning of the tryptophan operon in *E. coli* (9). The termination sites of RNA synthesis are less well explored but here also the general character (i.e., rich in uridylate residues) of sequences in the termination region has been recognized for several RNA species produced inside the cell as well as by RNA polymerase *in vitro* (10). In addition recent examination of transcription of the tryptophan operon has revealed a commonly used early termination site termed an "attenuator" (9). Even in times of severe tryptophan starvation, RNA polymerase molecules only transcribe through this region, which lies 166 nucleotides from the origin of transcription, about 50% of the time. These results focus attention on the importance which regulated termination can have on determining the boundaries of the transcriptional unit.

In eukaryotic cells no genetic definition of transcription units can be expected soon and without genetics to aid in establishing the functional limits of the DNA which contains a transcriptional unit for mRNA, a description of the unmodified primary RNA transcript would appear to be the most direct means of identifying a transcription unit for mRNA production. The first transcriptional unit to be recognized in eukaryotic cells was for ribosomal RNA (11,12) and this was initially identified during pulse labeling studies with RNA precursors. The basis for recognition of the rRNA transcriptional unit was the existence of a sufficiently stable large molecule with chemical similarity to ribosomal RNA that became labeled prior to the ribosomal RNA subunits themselves. Thus the idea of the transcriptional unit producing an RNA precursor molecule came from studies on eukaryotic cells. It is now widely recognized that in both bacteria and in eukaryotic cells this general rule applies to RNA formed from ribosomal and tRNA genes (13,15). In bacteriophage T7 replication a large transcription unit which produces an mRNA precursor molecule for several smaller specific mRNAs has also been described (14).

The ribosomal precursor RNA constitutes such a large proportion of the RNA formed by growing mammalian cells, it could easily be recognized by pulse-labeling techniques. Attempts to define the transcription units for mRNA molecules by the same pulse-labeling techniques suffer the obvious disadvantage that cultured cells make thousands of different mRNA molecules and thus at any one time many different transcription units contribute newly labeled RNA. It has therefore only been possible to demonstrate that a large amount of the non-nuclelolar RNA (the so-called hnRNA, heterogeneous nuclear RNA) generated in the cell nucleus is in a higher molecular weight form than that which is found as mRNA in the cytoplasm (15). Further, as mentioned above, the presence in hnRNA of terminal, post-transcriptionally-added sequences that are characteristic of mRNA is suggestive that the transcriptional units in eukaryotic cells which are responsible for mRNA production may generally be larger than mRNA molecules. Many questions about this proposed pathway of mRNA biogenesis can only be answered by measuring the transcription products of individual genes. For example, is the transcription unit for a given gene larger than the mRNA for that gene? Does the transcription unit of a given gene have precisely defined termini? Even in one cell type does a single transcription unit always give rise to the same product? During different stages of differentiation does one transcription unit give rise to one product?

New biochemical methods, notably those involving recombinant DNA, now promise answers to some of these questions in the next few years. Prior to the widespread availability of cloned mammalian cell DNA, cells infected with and transformed by DNA viruses provide the opportunity to clearly define at least some transcription units responsible for mRNA output. We have felt that such experiments would be important to provide a framework for future investigations that can define and follow the output of many more cellular transcription units. We have therefore returned to a study of transcription from adenovirus DNA late during infection (16,17,18). This virus contains a single linear DNA molecule of 35,000 base pairs (35 KB = kilo bases). In the cell, the DNA enters the nucleus where transcription ultimately results in the production of cytoplasmically translated mRNA molecules. The adenovirus mRNA resembles in every way the messenger RNA of uninfected cells: the size range is 1-4 KB (19), the 5' termini are blocked, methylated oligonucleotides, i.e. "caps" (20), and the 3' termini are segments of poly(A) (21). In addition to these identifying marks of mRNA, AD-2 mRNA's also contain internal methylations, principally N-6 methyl adenylic acid (m^6Ap). Thus far only mRNA molecules which come from the nucleus of the cell and enter the cytoplasm to be translated as mRNA bear this internal methylation (1). For example, the mRNA produced from vaccinia virus DNA in the cytoplasm contains a "cap" and poly(A) but lacks m^6Ap (22).

The study of transcription with pulse labeling techniques in adenovirus-infected cells late in virus infection offers the particular advantage that perhaps as much as 30% of the total RNA synthesis is adenovirus-specific RNA (23). In addition, high molecular weight RNA molecules considerably larger than adenovirus specific mRNA have been identified (24,25,26). A number of messenger RNA's have been demonstrated to be transcribed from right to left on the conventional physical map (Ref. 27, Fig. 1). Philipson and his colleagues (23) suggested that the majority of transcription late in infection was from the rightward reading strand. We have confirmed this conclusion by hybridizing pulse-labeled RNA to separated strands of AD-2 DNA. Over 98% of the adenovirus specific RNA labeled during a two-minute pulse hybridized to the rightward reading strand (Table 1) from which several molecules originate. Thus a determination should be possible of which region or regions of a specific strand of a single type of DNA produce the transcripts that ultimate lead to mRNA. The basic thrust of the experiments from our laboratory attempting to define the late AD-2 transcripts has been to examine RNA molecules

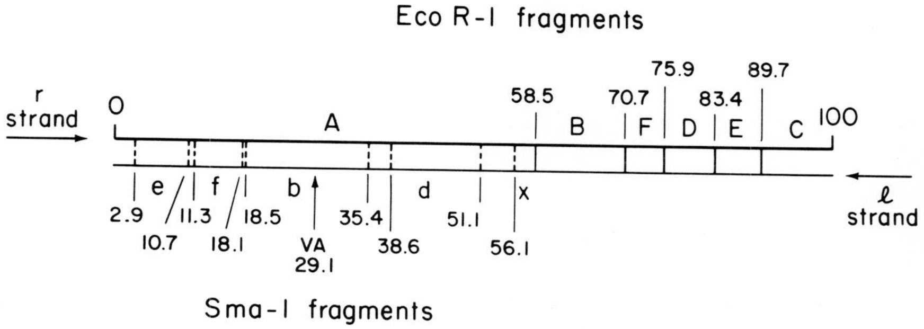

Fig. 1. Physical map of AD-2 DNA showing location of cleavage sites for Eco F-1 and Sma R-1 restriction enzymes. Transcription late in infection is mainly from left to right on the "rightward" reading strand, r strand. The cleavage sites are from Mulder et al., 1974 (36) and from a compilation of data by Dr. Marc Zabeau which was distributed at 1976 Cold Spring Harbor DNA Tumor Virus Workshop.

TABLE I

Strand Specificity of Transcription Late in AD-2 Infection

Exp.		Hybrid r	to l	Ratio r/l	Efficiency r/l	Corr. ratio r/l
1	^3H, 2 min. in vivo	11,000	80			
2		9,500	130	100/1		100/1÷
3		12,880	119	(avg.)		1.6 =
4		8,400	100			60/1
5	^{32}P, in vitro	6,450	4,250	1.5		
1	^{32}P (1/30 of Exp.5)	360	120	3.0	2.5÷1.5=	
2	^{32}P (" " ")	240	120	2.0	1.6	

^3H uridine labeled RNA was prepared from nuclei of AD-2 infected cells that had been labeled for 2 minutes beginning 15 hours after infection.

^{32}P RNA was prepared by in vitro transcription of AD-2 DNA by E. coli RNA polymerases which produces labeled RNA from both strands of the DNA. The ratio of ^3H labeled r strand hybrid to l strand hybrid ("efficiency") was used to correct the effect of the ^3H RNA on the hybridization of the ^{32}P RNA in order to obtain the corrected ratio of r/l.

in the process of their formation. Such molecules, "nascent" RNA or not yet completed RNA, should afford the best chance of observing newly formed molecules before they have become engaged in any possible post-transcriptional processing. Both inside the cell and in isolated nuclei the labeling time for RNA synthesis can be restricted to such a short period that only the termini of nascent molecules should have become labeled. Thus if a single rightward-reading starting point on the 35 KB of adenovirus DNA were responsible for most transcription, then the nascent molecules should form an ordered series such that the longest labeled molecules would have labeled RNA segments complementary to the right-most portion of the genome (Fig. 2). In contrast, if transcription

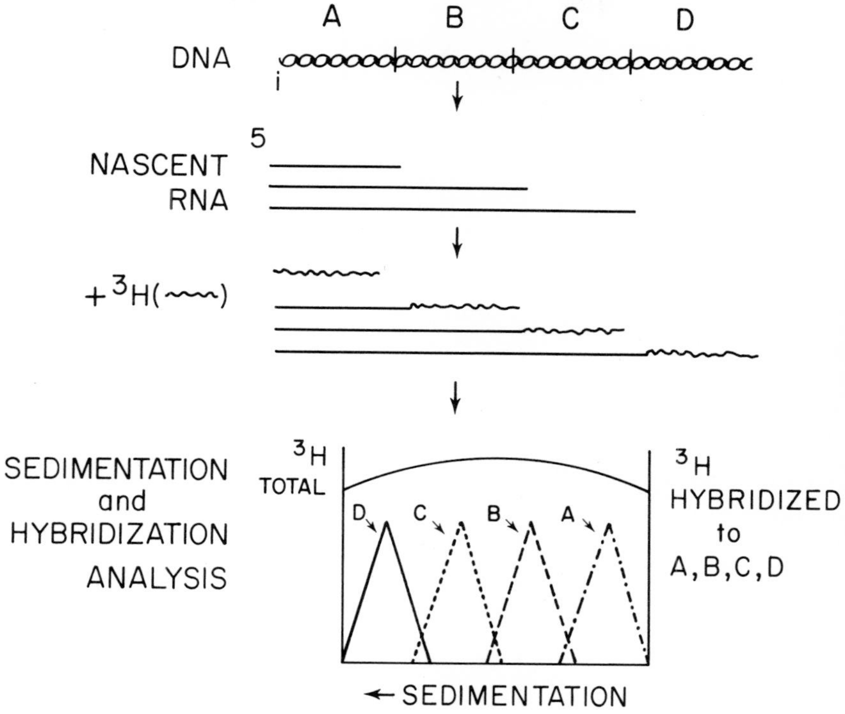

Fig. 2. Diagram of pulse-label experiment to detect, by sucrose gradient analysis, nascent RNA complementary to restriction endonuclease fragments A,B,C, and D of DNA.

were initiated independently for each relatively short mRNA molecule, then shorter nascent RNA should hybridize to virtually all regions of the AD-2 DNA. These ideas were tested by separating into size classes pulse-labeled RNA from labeled cells or from isolated nuclei labeled *in vitro* (16). The RNA was then hybridized to a series of DNA fragments created by restriction enzymes. The summary of the results obtained in this type of experiment are diagrammed in Figures 3 and 4 and Table 2. The longest labeled nascent molecules

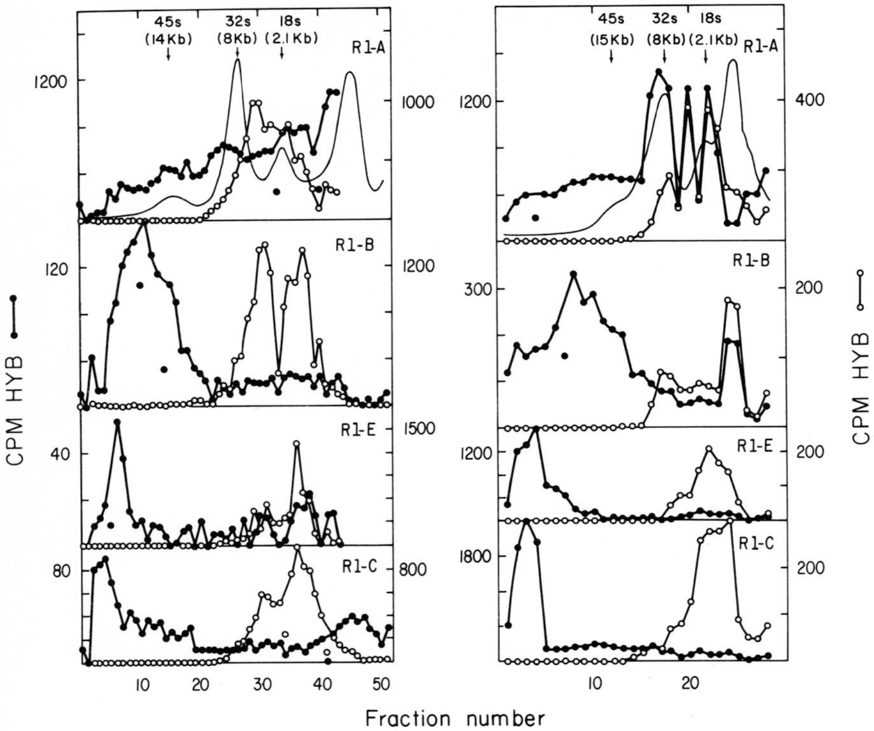

Fig. 3. Hybridization of pulse-labeled nuclear RNA with restriction fragments of Ad-2 DNA. Cells were labeled for 1 min. (left) or 2 min. (right) with (^{3}H) uridine (30 Ci/mmol., 200 ci/ml) at 18 hr. after infection, and nuclear RNA was isolated. (^{3}H) RNA was mixed with ^{32}P-labeled poly(A)-containing cytoplasmic RNA, isolated late in AD-2 infection, and centrifuged. Aliquots of individual fractions were hybridized with <u>Eco</u>R1-generated Ad-2 DNA fragments. Solid circles -o- pulse labeled RNA; open circles -o- ^{32}P-labeled cytoplasmic RNA; solid line A$_{260}$ absprbamce profile.

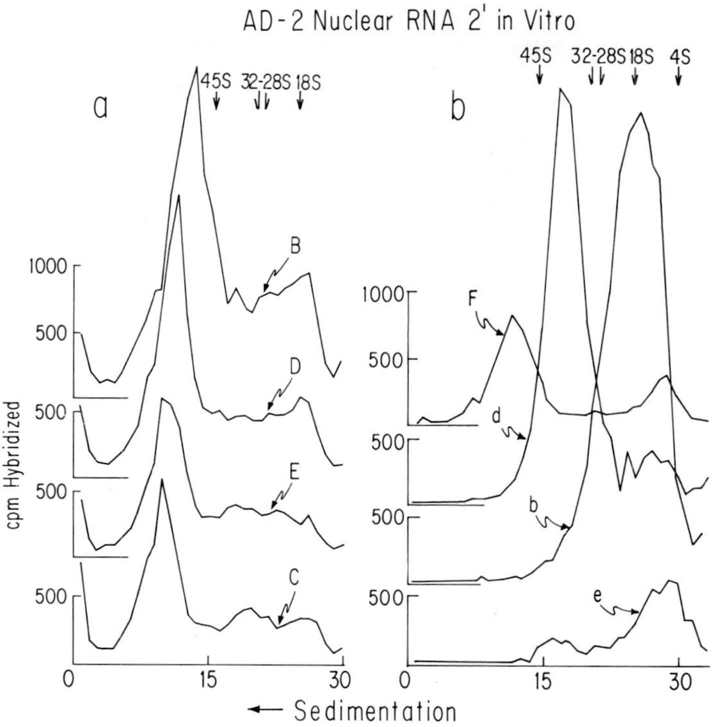

Fig. 4. Sedimentation distribution of RNA specific for AD-2 restriction fragments.

RNA labeled by a 2 minute exposure of unlabeled nuclei to ^{3}H-UTP was extracted, denatured and sedimented. Equal samples from each gradient fraction were hybridized to Eco R1 restriction fragments B,D,E, and C and (left panel), a second RNA preparation was hybridized to Eco F and Sma e,d, and b fragments (right panel).

TABLE II

Hybridization of "small" nuclear RNA

DNA fragment		CPM hybridized	% of total hybridized
Sma	e	247	5.
	f	1482	33.
	b	1859	41.
	d	321	7.
	x	32	.7
Eco	B	186	4.
	F	129	2.8
	D	120	2.6
	E	59	1.2
	C	115	2.5

RNA labeled in nuclei isolated from infected cells and the most slowly sedimenting one-third of the molecules (see Fig. 4) collected and hybridized to filters bearing AD-2 DNA fragments.

were complementary to the rightmost AD-2 DNA fragment, Eco C, and in the genome order, the next longest molecules hybridized to Eco E and so on all the way back to Sma b which lies from 0.18 to 0.35 on the adenovirus genome. Furthermore much as 90% (and at least 60%) of the transcriptional output of any region of the adenovirus genome from Sma b all the way to the right end of the genome could be detected as part of a single sedimenting species.

The transcriptional units for a variety of individual RNAs have been defined by a single sedimenting peak on sucrose gradients e.g., r-pre-RNA, (27a) polio RNA, (27b) silk fibroin mRNA, (27c) and "75S" Chironomous tentans salivary gland RNA (27d). The most plausible explanation of the result on nascent AD-2 specific RNA is that a single promoter exists in the region of ∿0.2; the RNA molecules which arise from the righthand ∿80% of the genome therefore all start at the same or very closely spaced initiation sites. Several interesting features of this putative initiation site are apparent from a more quantitative assessment of the data. Sedimentation size of the RNA complementary to Eco B, Sma b, i and d (17 and Weber, unpublished) indicate nascent chain lengths which appear to have been begun at a position approximately 0.2 on the genome. That is, the distances from 0.2 to the end, for example, of Sma b, i, and d or Eco B are equal to the chain lengths of the nascent labeled RNA complementary to these DNA fragments. Thus such a transcript would "read through" (i.e. contain) sequences for VA RNA, a small independently synthesized RNA molecules made by RNA polymerase III which is located at 0.29 (28,29). Further it appears that most, if not all, of the polymerase molecules which initiate in the region of 0.2, transcribe all the way to the end of the genome because there is approximately equimolar transcription from all DNA fragments whether the label was two minutes in vitro or 2 to 5 minutes in vivo (17). Finally it appears that polymerase II which earlier had been demonstrated to be responsible for the majority of RNA synthesis late in adenovirus infection (30,31), is indeed responsible for the synthesis of the RNA molecules observed in the sedimentation profiles: all of the RNA peaks observed in Figs. 2 and 3 are inhibited by the inclusion of α amanitin in the isolated nuclear incubation mixture (Weber, unpublished).

Because approximately 20 to 40 percent of the RNA transcribed in the pulse label was not part of a single dominant peak, the possibility remained that the immediate messenger precursor molecules might still be short transcripts, although such short transcripts of a specific size did not appear to be an important part of the transcription pattern. We therefore sought independent means of determining whether transcription for the righthand portion of the adeno genome was dependent exclusively on distant promoters. The technique of Saurbier and colleagues (32-35), termed UV transcription mapping, has been applied to adeno-2 RNA synthesis (18). The theory of these experiments is that the thymine: thymine dimers found by UV radiation cause premature chain termination, but no other lesion in RNA transcription. Polymerases thus start and stop prematurely on the UV damaged template but start again in the correct position. Thus RNA synthesis in a given transcription unit is decreased within the unit as an exponential function of distance from the promotor. UV transcriptional mapping reached the correct conclusions that bacterial ribosomal RNA synthesis and early T7 messenger RNA synthesis in E. coli originated within single transcription units (33,34). Also UV transcription mapping successfully predicted the correct sequence order (18S→28S, 5'→3') as well as the transcription unit size for ribosomal RNA in L cells (35).

HeLa cells late in AD-2 infection were exposed to UV irradiation and assayed for the effect of transcription on specific regions of the AD-2 genome. Pulse labeled RNA from control or UV irradiated cells was hybridized to various segments of the AD-2 genome and the percent of RNA transcription which survived was plotted on a logarithymic scale (Fig. 5) as a function of location on the linear

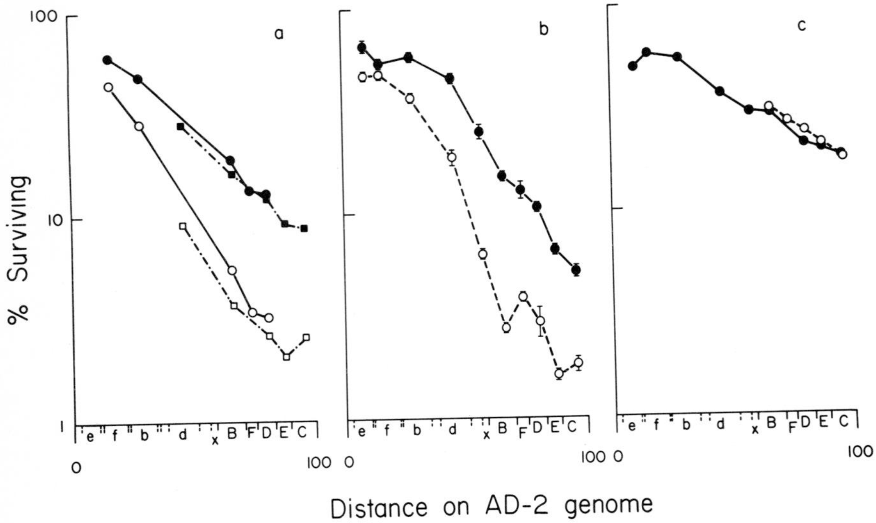

Fig. 5. UV inactivation of nuclear RNA synthesis from various regions of the AD-2 genome.

Nuclear RNA from control and UV irradiated AD-2 infected cells was prepared and hybridized to the indicated AD-2 DNA fragments immobilized on nitrocellulose filters. The amount of hybrid to each DNA fragment in the UV samples was divided by the control for that region x 100 to give the "% survival" plotted in the graph. Three different experiments are shown; the upper and lower curves represent two different UV doses in individual experiments a) 25 (-o- and - -) and 50 seconds of UV (-o- and - -; the two sets of symbols indicate two separate hybridizations of RNA to DNA fragments), (b) 35 (-o-) and 70 (-o-) seconds of UV and c) 25 seconds (the two symbols represent two separate hybridization of RNA to DNA fragments).

AD-2 DNA molecule. An exponential decrease was observed for RNA transcription as a function of genome position with the rightmost region of the AD-2 genome being most sensitive. Such an exponential decrease is consistent with a major promotor in the region of about 0.2 to 0.3 that is responsible for the transcription of the righthand 70-80% of the adenovirus genome. Furthermore, it appears for the *Eco* E and C fragments that 95% of the RNA must derive from such a distant promotor, because the UV-induced decrease in transcription remains exponential through a survival of 5% for these fragments.

From the pulse-labeled experiments plus the UV experiments, it seems likely that in excess of 90% of the transcription which occurs from AD-2 DNA late in infection begins somewhere in the region of 0.2 and reaches to the righthand end of the genome. Since at least 20% of the RNA formed in the nucleus exits to the cytoplasm (recent unpublished experiments by J. Nevins suggests as much as 30 to 40% conservation may occur for those regions of the genome that are productive for messenger RNA), we conclude that the cell possesses the capacities for synthesizing a long RNA molecule which is then conducted stepwise through to messenger RNA formation. The steps must include cleavage to develop both a 5' and 3' end, and posttranscriptional addition of a cap to the new 5' end, and of a poly(A) segment to the new 3' end.

CONCLUSION

These studies contribute information to the details of messenger RNA manufacture in adenovirus infection. But perhaps more significantly, they demonstrate a range of approaches to define a transcriptional unit in mammalian cells. For example, pulse-labeling both in vivo and in vitro has allowed an observation of the same nascent molecules prior to their cleavage. Isolated nuclei appear to elongate the same molecules that were in the process of being formed inside the cell. If such elongation of nascent RNA occurs to all chains equally, and this appears to be the case at least in AD-2 infected cells, the labeling of isolated nuclei affords an opportunity to measure the instantaneous rate of any particular RNA sequence relative to the total. Such a measurement will obviously be critical in the solution of the locus of regulation of gene expression in the cellular genome. Thus when plasmids containing DNA complementary to cell DNA become available, studies patterned exactly along the lines of the study of the transcription of the AD-2 genome should be most informative. Transcription units can be identified and their output measured relative to the utilization of their output as mRNA.

REFERENCES

(1) A. J. Shatkin. New Scientist, in press. (1977).

(2) J. E. Darnell, W. R. Jelinek and G. R. Molloy. Science 181 (1973) 1215.

(3) R. P. Perry, D. E. Kelley, K. Frederici and P. Rottman. Cell 4 (1975) 387.

(4) M. Salditt-Georgiev, W. Jelinek, J. E. Darnell, Y. Furuichi, M. Morgan and A. Shatkin. Cell 7 (1976) 227.

(5) R. H. Herman, J. G. Williams and S. Penman. Cell (1976) 429.

(6) R. A. Steinberg, B. B. Levinson and G. M.

Tomkins. Proc. Nat'l. Acad. Sci. U.S. 72 (1975) 2007.

(7) W. Gilbert, N. Maizel, and A. Maxam. C.S.H. Symp. Quant. Biol. 38 (1973) 845.

(8) T. Maniatis, M. Ptashne, K. Backman, D. Kleid, S. Flashman, A. Jeffrey and R. Maurer. Cell 5 (1975) 109.

(9) K. Bertrand, C. Squires and C. Yanofsky. J. Mol. Biol. 103 (1976) 319.

(10) J. Roberts. Proc. Nat'l. Acad. Sci. U.S. 72 (1975) 3300.

(11) K. Scherrer and J. E. Darnell. Biochem. Biophys. Res. Comm. 7 (1962) 486.

(12) K. Scherrer, H. Latham, and J. E. Darnell. Proc. Nat'l. Acad. Sci. U.S. 49 (1963) 240.

(13) J. Smith. Prog. Nuc. Acid Res. 16 (1975) 25.

(14) J. J. Dunn and W. Studier. Proc. Nat'l. Acad. Sci. U.S. 70 (1973) 3296.

(15) J. E. Darnell. Harvey Lectures 69 (1975) 1.

(16) S. Bachenheimer and J. E. Darnell. Proc. Nat'l. Acad. Sci. U.S. 72 (1975) 4445.

(17) J. Weber, W. Jelinek and J. E. Darnell. Cell (1977) in press.

(18) S. Goldberg, J. Weber and J. E. Darnell. Cell (1977) in press.

(19) U. Lindberg, T. Persson and L. Philipson. J. Virol. 10 (1972) 909.

(20) S. Sommers, M. Salditt-Georgieff, S. Bachenheimer, J. E. Darnell, Y. Furuichi, M. Morgan and A. J. Shatkin. Nuc. Acid Res. 3 (1976) 749.

(21) L. Philipson, R. Wall, R. Glickman and J. E. Darnell. Proc. Nat'l. Acad. Sci. U.S.A. 68 (1971) 2806.

(22) B. Moss. Proc. Nuc. Acid Res. 19 (1977) in press.

(23) L. Philipson, U. Petterson, U. Lindberg, C. Tibbetts, B. Vennstrom and T. Persson in : Cold Spring Harbor Symp. Quant. Biol. 34 (1974) 447.

(24) P. M. McQuire, C. Swart and L. D. Hodge. Proc. Nat'l. Acad. Sci. U.S.A. 69 (1972) 1578.

(25) R. Wall, L. Philipson and J. E. Darnell. Virology 50 (1972) 27.

(26) J. T. Parsons and M. Green. Virology 45 (1971) 154.

(27) P. A. Sharp, P. H. Gallimore, and S. J. Flint. Cold Spring Harbor Symp. Quant. Biol. 39 (1974) 457.

(27a) H. Greenberg and S. Penman. J. Mol. Biol. 21 (1966) 527.

(27b) J. E. Darnell, M. Girard, D. Baltimore, D. F. Summers and J. F. Maizel in: Molecular Biology of Viruses, ed. J. Colter (Academic Press, New York 1967) 375.

(27c) P. Lizardi. Cell 7 (1976) 239.

(27d) B. Daneholt. Cell 4 (1974) 1.

(28) U. Petersson and L. Philipson. Cell 7 (1975) 1.

(29) M. B. Matthews. Cell 6 (1975) 223.

(30) R. Price and S. Penman. J. Virology 9 (1972) 621.

(31) R. D. Wallace and J. Kates. J. Virology 9 (1972) 627.

(32) W. Sauerbier and A. Brautigam. Biochem. Biophys. Acta 199 (1970) 36.

(33) W. Sauerbier, R. L. Millette and P. B. Hackett. Biochem. Biophys Acta 199 (1970) 209.

34) P. B. Hackett and W. Sauerbier. Nature 251 (1974) 639.

(35) P. B. Hackett and W. Sauerbier. J. Mol. Biol. 91 (1975) 235.

(36) C. Mulder, J. R. Arrand, H. Delius, W. Keller, U.

Petersson, R. J. Roberts and P. A. Sharp. Cold Spring Harbor Symp. Quant. Biol. 39 (1974) 397.

Discussion

N. Chiu, National Institutes of Health: When you study the RNA synthesis with nuclei from adenovirus infected cells, how do you prepare the nuclei so that you can avoid the ribonuclease contamination after you break up the cell?

J. Darnell, Rockefeller University: Well, the question would be more pertinent if we were not getting the high molecular weight molecules. We are able to observe high molecular weight molecules which are DMSO stable, probably due to just simple good fortune that Hela cell extracts have relatively little nuclease activity. Thus, one is able to get from them intact strands of RNA that are in excess of 20,000 nucleotides. This was first observed years ago when Hela cells were used as the first cultured cells to be examined for polysomes.

D. Roufa, Kansas State University: Did you say that the high molecular weight transcripts are decorated at both ends, methylated and 3'-polyadenylated?

J. Darnell: Yes, the general results that have come out of Bob Perry's laboratory and ours, are that there are poly-terminated nuclear RNA molecules which do have caps on them. Even if you take the polyterminated units that exceed say 5 or 7,000 nucleotides in size you can still find caps. Now, that does not prove, of course, that either end of these molecules exists in the cytoplasm. If, however, the two ends of such a molecule do not then the pathway for cellular molecules could be analogous to what apparently happens to an adenovirus precursor molecule.

P. Duesberg, University of California, Berkeley: What is the relationship between the complexity and the size of these large RNAs like the hemoglobin RNA or adenovirus RNA? Do you think they all contain everything between the promotoride or purity promotocide in the end, or are they redundant transcripts of perhaps the same sequence?

J. Darnell: We can give you a fairly direct answer for adenovirus because we have hybridized all of the fragments between .2 and 1.0 and there it appears that everything exists in an equimolar fashion. In the case of the hemoglobin I do not know that it has been analyzed yet.

J. Hassell, Cold Spring Harbor Laboratory: Have you looked at the size of the adenospecific transcripts in the nucleus at early times?

J. Darnell: Yes, we have. Steven Bachenheimer re-examined that question recently and could not find evidence of large transcripts early in infection using the same techniques with which large molecules were easily found late in infection.

J. Hassell: Do you have any idea whether or not the template for late transcription is single-stranded DNA or double-stranded DNA?

J. Darnell: No, I do not; that is a good point, and we do not know the answer.

C. Wei, National Institutes of Health: I would like to make one small comment on your statement on the internal methylation. You mentioned that internal methylation is always present when the mRNA is derived from the nucleus.

J. Darnell: Well, I was under the impression that Moss believes that there is no m6AP in the vacinia mRNA - at any rate there is far far less than one per molecule. And there are many messengers in the cell that have not only one per molecule but several per mole. In addition influenza virus mRNAs, about which there has been speculation for years, now also has been found to contain m6AP to cytoplasmic mRNA. In general, it is the hallmark of the message that has passed through the nucleus and exited to the cytoplasm.

PANEL DISCUSSION

THE SOCIAL IMPLICATIONS OF RESEARCH ON GENETIC MANIPULATION

CHAIRMEN: DeWitt Stetten, Jr., National Institutes of Health
Julius Schultz, Papanicolaou Cancer Research Institute

PANELISTS: A. Bayev, U.S.S.R. Academy of Sciences
Paul Berg, Stanford University
Bernard Davis, Harvard University
Maxine Singer, National Institutes of Health
Robert Sinsheimer, California Institute of Technology
John Tooze, European Molecular Biology Organization at Heidelberg

Dr. Schultz:

Recent articles featured within the *New York Times*, *National Geographic*, and the *Wall Street Journal*, as well as programming by the prime television networks, have aroused intense public interest in recombinant DNA research. In response to this interest the Papanicolaou Cancer Research Institute has chosen to devote their final session of the Ninth Miami Winter Symposia to a panel discussion of the topic, "Social Implications of Research on Genetic Manipulation".

We have assembled a panel of outstanding spokesmen to discuss the issue. Not only have the panel members assisted in the formulation of the Guidelines established by the NIH, but they have also been active in explaining them to the public through appearances before Congressional Committees. They are not all in complete agreement, but each has given a great deal of thought to the subject. The opportunity to present their views to an informed and knowledgeable audience is an unusual one, particularly because this audience is composed of those scientists whose future may depend upon the panel's success in attracting adequate public support for their views.

I will now turn the meeting over to Dr. DeWitt Stetten who is the Deputy Director for Science at NIH, and who is also the Chairman of the Committee for determining the

Guidelines for protection against the biohazards created by recombinant DNA research.

Dr. Stetten:

Dr. Schultz, ladies and gentlemen, allow me to introduce the other members of the panel: Dr. Paul Berg of Stanford; Dr. Bernard Davis of Harvard; Dr. Maxine Singer of the National Institutes of Health; Dr. Robert Sinsheimer of the California Institute of Technology; Dr. John Tooze of the European Molecular Biology Organization at Heidelberg; and Dr. A. Bayev, U.S.S.R. Academy of Sciences, Moscow.

The program will consist of individual presentations by the panel members, then a free period when each panel member will be invited to ask a question of one other panel member, then some written questions which members of the audience have submitted (and which Dr. Singer has reviewed), and finally, free questions from the floor as our time permits.

I have obtained Dr. Singer's permission to open the proceedings with a small parable which I tried out on her a few days ago. The parable relates not to recombination of DNA molecules, but rather to the more general matter of genetic recombination - an event which occurs and has occurred for millennia on this planet. It is of course the essence of sexual reproduction, but it also occurs between bacterial forms, between viruses and bacteria, and there is evidence to indicate that it may be occurring between bacterial forms and eukaryotic cells.

A popular form of experimentation in this area results in the fertilization of the human ovum. Under these circumstances the ovum may be pictured as the host (to borrow from contemporary jargon), and the spermatozoa clearly becomes the vector bearing a charge of DNA which is introduced into the host where recombination occurs. The reason for the popularity of this experiment is because it offers enormous promise of useful outcome. It is, however, an experiment not devoid of hazard. In this case the hazard is not a potential hazard, but a very real and well-documented one. Textbooks of obstetrics contain fascinating pictures of some of the monsters which have resulted from this process, and even there what appear to me to be the major monsters are not presented. I refer to the societal monsters like Al Capone, or the international monsters like Adolf Hitler, who were clearly the result of such experimentation. I would point out that, at least in the case of the "gangster monster", he clearly has a survival advantage over me. All I have to note is the direction in which his machine gun is pointed.

I would now like to direct your attention to the experiment conducted by Adolf Hitler's parents about 90 years ago which resulted in the recombination of genetic information which was subsequently cloned, as you will recall, and resulted in the development of a situation which ultimately led to the premature death of approximately forty million people. I mention this to indicate the magnitude of the hazards of this particular experiment.

As the result of an assessment of this hazard, a Presidential Commission has been established, under the guidance of the Vice President, whose function will be to declare a moratorium on such experimentation in the future and to consider how this field of knowledge may be best managed. It is, I believe, their recommendation that cautious and judicious experimentation should be resumed, but under P4 conditions and in Class III safety cabinets, whose exit-ports are guarded by incinerators or autoclaves. Only in this way can we be quite certain that the monsters which may result from this experiment will not escape into and pollute the environment. Only in this way can we be reassured that the eco-niche which we so comfortably occupy will not be assumed by another generation who surely would not occupy it as nobly as we have done.

Our first speaker will be Dr. Robert Sinsheimer.

INDIVIDUAL PRESENTATIONS

Dr. Sinsheimer:

Thank you Dr. Stetten. I might note that the one saving grace to the experimentation that Dr. Stetten was describing is that the products all ultimately die out.

The recombinant DNA issue has become a meeting place for old rancours and new fears. It has become the testing ground for differing concepts of scientific responsibility. I hope that we can continue to explore our differences in a civil and objective manner (as becomes scholars and scientists), and that we may all recognize that monopolies on wisdom are rather rare.

I recently read an article by Milton Mayer in which he discusses dialogue. He says that dialogue is "not discussion or debate; it is not a talk show nor a brain display of expertise, or esoterica, or vanity; it is not a demonstration or an exchange of information. It is a cooperative inquiry directed at: 1) the formulation of a problem which is not susceptible to convincing empirical demonstration; 2) the statement of its possible solutions; and 3) the methodical consideration of those solutions in terms of both common and particular circumstances. It is an instrument for continuing investigation of matters which lie outside the realm of positive truth." I would say that according to this definition the recombinant DNA issue certainly qualifies as a fit subject for dialogue and, as a contribution to such a dialogue, I would like to make a few comments today on several aspects of this issue.

I realize that after you have listened for most of the week to the elegant science that can be, and is being, done with the aid of recombinant DNA techniques, you are not likely to be very receptive to suggestions that such work is potentially dangerous and should be subjected to increased constraints. However, it is now good form to include one session of contrition like this to demonstrate our sense of social responsibility and then, ultimately of course, to reaffirm the pledge of allegiance to the NIH Guidelines and reassert our confidence in P2 and EK2, the safe vehicle self-destructing with new clones and plasmids for all!

All is well. The ringleaders of recombinant DNA are hale and hearty as you can see. The morbidity in Palo Alto is no higher than usual and it seems safe to let the heretics rehearse their slightly "travel-worn" arguments and alarms. Few are likely to be persuaded. Let them ramble about P4 and _E. coli_. Recombinants are surely already abroad in the land. P3 laboratories are springing up like mushrooms (the symbolism is intentional) and who is the worst. Well, I submit that unfortunately we, the biology community, are the worst and that we are perceived from outside to be (in hyperbole) holding the people of America captive while we perform an experiment upon them. We are perceived to be subjecting society to an unquantifiable risk on the basis of our work that such risk is small and justified by our definition of the "greater good". Now this action may be seen as no worse than those of the nuclear industry, or the oil industry, or the chemical industry - but we were formally perceived as different.

I have discussed this subject with many groups at universities and outside academia, and I find, to my dismay, that we are acquiring an unfamiliar image. In this image it is perceived that scientists too can cohere into yet another parochial establishment - jealous of its unquestionably worthy prerogatives, resistant to institutional change, attached to convenience and unmoved by less than immediate concerns. Now let me say that I do not believe this is a true image. I am trying to report something to you. Of course some of you may feel that this perception exists only because Bob Sinsheimer and a few like him are "whipping on" the populace. Do not be so sanguine - the populace needs no "whipping on".

There is, I regret to say, a deep-lying mistrust of science in general today and of modern biology in particular. There is a fear of what it may bring to our lives which boils up to the surface in Cambridge, San Diego, Princeton and Ann Arbor, which has been regrettably (and unintentionally I am sure) abetted by the way in which this particular issue has been handled. Hindsight is great of course. This concern is usually presented in rather simplistic fear-laden terms of calamity, but I believe these mask less immediate, but deeper lying, more poorly articulated concerns about morality, about individuality, about man-made change in the world of life, and about the qualities of a humane existence.

So far the dialogue over recombinant DNA (the attempt to secure the benefits of this new technology without exposing ourselves to new and intractable torments at the same time) is viewed as a portent of things to come. The manner of conduct of this issue has not inspired great confidence that future issues will be wisely met in a democratic manner. In the popular mind that I have met, the nature of these future concerns is rather amorphous based on a hint here, and a guess there, and frankly it is rather confused. Thus cloning is confused with recombinant DNA, and human genetic engineering is viewed as an imminent problem. But more cogently, and quite realistically among more informed people, it is not hard to envision a long procession of future issues of similar gravity. Can we explore a therapy for schizophrenia without the risk of discovering easily produced new opiants? Can we explore a cure for cancer without the risk of discovering simple means for major extension of the human life span with attendant social chaos? These are not wholly imaginary concerns, and how will they be approached?

Now I know some believe that we have no real choice - that science is the proverbial irresistible force and that what will be, will be. But, remarkably I do not believe society is willing to accept such a fatalism. The particular arguments over recombinant DNA are now well known and need hardly be repeated here. They have been debated on a dozen campuses and communities. The summation of each debate generally goes about like this: "The benefits of this technology are uncertain, but probably feasible; the hazards of this technology are undemonstrated and probably small; the Guidelines provide a prudent compromise or balance between benefit and risk; so - let us proceed. Besides, if we don't our scientists will move elsewhere." To a sociologist all this must be fascinating. As a participant I find it dismaying. The last part is certainly true; it demonstrates the impracticality of coping with this problem on a local basis as has indeed already been demonstrated on a number of prior national issues. But I believe the earlier betrays a great loss. I must confess that I have not appreciated that moral relativism is so completely tried - even the language of morality has vanished. We frame the debate in benefit to risk ratios, without regard to whose benefit or whose risk. Does the calculus have the accountability to weigh benefits and risks equally? One wonders if we have all somehow just become numb; if the daily threat to life inherent in the nuclear age has somehow coarsened us; if the sheer abundance of humanity has subtly changed its values, or if even the specter of epidemic pales in the light of a hundred million human births per year - the kind that Dr. Stetten was describing.

I would like to comment briefly (and I do this not in any sense of trying to cast any blame) on the process in which the discussion has been conducted and by which the Guidelines were developed for I believe therein lies the root of the problem I have described. We seem to view these issues, such as recombinant DNA, in isolation. We do not see their antecedents and their consequences, nor their inevitable connections to the entire societal ambiance. We do not see their prospective roles as precedent and guide for the conduct of future discussions. We do not see them as experiments from which progressively to learn how to go about the resolution of these complex matters. Instead each issue, indeed each proposal, seems to become a contest - not an exploration; a zero sum gained to be won or lost instead of a creation. The issues are not approached with a sense of community; with the recognition that in fact we all share the hazard; with the perception that we are always part of a larger society whose concerns, even if less well-informed, do have validity and should be either allayed or accepted.

I believe there was a set of basic and prior policy issues which needed resolution before guidelines were developed. The absence of such resolution frustrated this dialogue.

We confront here for the first time the question of regulation of a sector of basic science. Among others, we need policy judgments with respect to such questions as: To what extent can and should scientific research be regulated and for what purpose? Can such regulation be self-regulation or is the force of law a necessity? On whom should lie the burden of proof? Do those who wish to experiment have to prove their research will not be dangerous?

We have, of course, recently witnessed similar debates concerning the drug industry, the chemical industry, and the nuclear industry - but each is a different case. Should, and can, a clear distinction be made between agencies which sponsor, and agencies which regulate? How do we evalute and take into account social hazards as well as technical hazards? How, and to what extent, should the public participate in decisions to regulate science and in the practice of such regulation, if any? I am sure that you could add to the list. Had these policies been worked through in a public forum before technical guidelines were formulated, I believe we would not now be witnessing the current intense and unsatisfied public concern. We debate technicalities over and over because we have not yet resolved the principles.

I believe that we urgently need to establish, as others have suggested, either a commission to consider these questions, or alternatively perhaps, to reconvene the broadly constituted ad hoc Advisory Committee to the Director of NIH, which considered this matter, but never actually had an opportunity to function as a committee. In this connection and with this issue there is a curious, and perhaps not widely recognized, inversion of ordinary assumptions. Science is an ultra-elite affair. In the conduct of scientific research we need be concerned really only with the activities of the most competent scientists in the best laboratories. As a self-correcting enterprise the presence, even the mistakes, of less competent members of the community are ultimately of small consequence. They neither contribute to, nor particularly impede, scientific progress. However, when we consider the potential for hazard the rules are inverted. It is the most careless and the least competent laboratories, the lowest common denominator, that needs the greatest concern.

There have been complaints that criticisms of the Guidelines have not been very specific. Of course, they have not and that is precisely the point. If the potential hazards were few and clearly defined, they could be analyzed and hopefully shown to be negligible and we could relax. But unfortunately,the number of potential scenarios of hazard is vast and varied and we lack the general theorems that would permit us to exclude with confidence all classes of hazard. Indeed we should remember that all of our theories, even in the more fully developed scientists, have their limits which we discover in good time as we extend the range of our observation and then develop more encompassing theories. But with recombinant DNA our practice far outpaces our theories and may carry us swiftly into new domains and new perils.

The proponents of the Guidelines frequently conclude with the cliche that there is no riskfree course of action. Well I agree. But the risks are various in different courses of action and subject different populations to different risks. My own view, which I know some regard as exaggerated and unwarranted, is that once again we are taking an unquantifiable and unnecessary risk of calamity on the basis of intuitive judgments. We do not need to advance our science by imposing upon society such decisions or the purity of our motives. To quote T.S. Eliot, "Neither fear nor virtue saves us". And, of course, Pogo said it best, "We've met the enemy and they is us". Thank you.

Dr. Stetten: Our next speaker is Dr. Bernard Davis of Harvard.

Dr. Davis:

Recombinant DNA research has led to extensive public discussions of two potential risks: the immediate risk of harm from some of the novel organisms produced, and the more conjectural, long-term risk that our interference with evolution will eventually create unforeseeable disasters. These discussions have been based largely on the assumption that any novel organism produced by this technique may well survive and spread. But this assumption ignores Darwin's great discovery: the dominating role of natural selection in determining what survives, multiplies, and evolves. While Darwin dealt only with the visible living world, Pasteur made essentially the same discovery for invisible organisms, though expressed in different terms: bacteria do not arise by spontaneous generation but are ubiquitous, and the kinds that grow out in any medium are the ones that are selected by that medium. An extension of these principles to infectious disease gave rise to the science of epidemiology, which may be viewed as a branch of microbial ecology concerned particularly with the distribution of pathogenic organisms.

I believe that epidemiology and evolution have been grossly under-represented in the professional as well as in the public discussions of the problem. My credentials for correcting this defect are modest, for my research shifted away from medical microbiology a good many years ago, and since then my contact with the field has depended mostly on textbook writing. But, I hope the information I offer will encourage deeper exploration of these topics by experts in future symposia.

To be sure, the most reliable basis for assessing the hazards will ultimately be provided by the accumulation of experience with recombinants. But meanwhile we should not act as though we are entering this new territory with no knowledge to guide us: we have a good deal of pertinent information from evolution and from epidemiology. To molecular biologists who have seen one deep mystery after another in other areas of biology settled by the extremely hard data that their field provides, the evolutionary considerations that I shall invoke may seem like mere handwaving. But in this light nearly all of Darwin's arguments, based on inferences about the past and not on verifiable experiments, could be similarly dismissed. And I would remind you that Darwin's theory remains the most profound and unifying generalization in biology: it is enormously supported today by the evidence from DNA sequences for the genetic origin and in the continuity of the observed variation, but it also involves ecological processes and populational kinetics that require a totally different set of concepts and approaches from those of molecular genetics.

Let us now review some of the pertinent principles from evolution and microbiology.

1) The Meaning of Species. As evolution created the process of sexual reproduction, whose reassortment of genes provides a vastly increased supply of genetic diversity for the mill of natural selection, it also developed species: groups of organisms that reproduce only by mating with other members of the same group, and not with members of other species. The evolutionary function of these fertility barriers is clear: diversity is necessary for evolution, but since a successful organism must have a reasonably balanced set of genes the diversity resulting from unlimited combinations from the pool of genetic material in the living world would not be useful. Species barriers eliminate the production of grossly unfit, nonviable progeny.

Unlike eukaryotes, prokaryotes ordinarily reproduce by the asexual process of cell division. Their occasional gene transfers do not show a sharp species boundary: the transfer simply becomes less efficient the greater the evolutionary separation between the donor and the recipient. Prokaryotes therefore have no true species: they have an almost continuous spectrum of genetic patterns, and borders between what we call bacterial species are arbitrary and often controversial. E. coli, for example, is the name given to a range of strains with certain common features but also with a variety of differences, and these differences determine their relative Darwinian fitness for various environments. This elementary concept was entirely missing from Cavaliere's discussion of the hazards of inserting genes in E. coli in the *New York Times Sunday Magazine* last August, and I would criticize the *New York Times* for publishing such a polemical and unqualified account.

2) Bacterial Ecology. Every living species is adapted to a given range of habitats. The set of bacterial strains called E. coli thrive only in the vertebrate gut, and because these cells die out rather quickly in water the E. coli count of a pond or a well is a reliable index of its continuing fecal contamination. In the gut there is intense Darwinian competition between strains, depending on such variables as growth rate, growth requirements, ability to scavenge traces of food, adherence to the gut linings and resistance to antimicrobial factors in the host. Hence most novel strains are quickly extinguished. The mechanism of this extinction is the kind of selection by competition envisaged by Darwin for higher organisms, but with bacteria it happens in days rather than in eons.

This effect of the environment in the gut on the normal flora is readily recognized. For example, when breast feeding is replaced by solid food that character of the stool changes dramatically, as lactic acid bacteria (which produce sweet-smelling products) are replaced by E. coli and other foul organisms. Early in this century Mechnikov romantically hoped to promote longevity by supplying lactic acid bacteria, in the form of yogurt, to displace the presumably toxic foul organisms. The experiments were a dismal failure, but the commercial success is still seen.

3) Pathogenesis. Only an incredibly small fraction of all bacterial species can cause disease. The rest play essential roles in the cycle of nature. Infectious bacteria differ from each other in several distinct respects:

infectivity (i.e., the infectious dose, ranges from a few cells of the tularemia bacillus to around 10^6 of the cholera vibrio); specific distribution in the body; virulence (i.e., the severity of the disease produced); and communicability from one individual host to another. These attributes depend on the coordinate activity of many genes, which are capable of independent variation. For our discussion the distinction between the ability to produce a serious disease and the ability to spread is particularly important.

4) Types of natural selection. When an organism grows continuously in a relatively constant environment natural selection has a stabilizing effect, weeding out the variants that deviate too far in any direction from the well-adapted norm. But when the environment is changed the same basic process of natural selection has a diversifying effect: the new circumstances select for the preferential survival and reproduction of variants with increased fitness for those circumstances. This Darwinian process explains a phenomenon that confused early workers: when pathogenic bacterial strains are isolated from infected hosts and then repeatedly transferred in artificial culture media they often rapidly lose virulence. We now know the mechanism by which this improved adaptation to the new environment occurs, at the expense of decreased adaptation to the old one (i.e., loss of virulence): the original strain is gradually outgrown by the progeny of rare mutants that are better adapted to the new culture medium (i.e., that can grow slightly faster, or can grow slightly longer with a limited food supply).

It is clear that natural selection plays an overwhelming role in evolution, though with bacteria its role was long unrecognized: the population shifts seemed too rapid for an undirected process, and the existence of genes and mutations in bacteria was not recognized until the 1940s. But by now selection has become the foundation of bacterial ecology.

With this background now, let me consider the hazards from the organisms that are being produced and will be produced. In trying to estimate the immediate hazards from novel organisms it is useful to distinguish these possibilities: that experiments with a given kind of DNA will produce a dangerous organism; that the organism will infect a laboratory worker; and that the organism will spread.

I would like to concentrate on a kind of experiment that is causing great concern and is restricted to quite special facilities: the so-called "shotgun" experiment with random fragments of DNA from animal cells. Two considerations seem pertinent. First, the probability that any fragment will contain a gene for a toxic product, or the genes of a tumor virus, is exceedingly low, though not zero. Second, evolutionary considerations provide an independent approach to the question. It seems exceedingly doubtful that our novel ability to introduce mammalian DNA into bacteria in the laboratory will create a truly novel class of organisms, for evolution has already had a crack at the problem.

In particular, it is known that bacteria can take up naked DNA from solution; and, in fact, transfer of DNA between two strains of pneumococcus has been demonstrated in the animal body. Moreover, bacteria in the gut are constantly exposed to fragments of host DNA that are released as the cells lining the gut die, and bacteria growing in carcasses have a veritable feast of DNA. The efficiency of such uptake of mammalian DNA by bacteria is undoubtedly very low. However, because of the extraordinarily large scale of the exposure in nature, recombinants of this general class must have been formed innumerable times over millions of years. They have thus been tested in the crucible of natural selection, and if they had high survival value we would be recognizing short stretches of mammalian DNA in E. coli. We do not. If, on the other hand, naturally occurring recombinants are appearing and even causing disease, but are escaping our attention, we would have to ask how much our laboratories could add, since nature experiments with about 10^{20} - 10^{22} bacterial cells produced in the human species per day.

Let us now consider the probability that an inadvertently produced harmful organism might cause a laboratory infection, and let us assume the worst case: an E. coli strain producing a potent toxin absorbable from the gut, such as botulinus toxin. Such a strain would indeed present a real danger of laboratory infection. But there are a number of reasons to expect this danger to be less than that with the pathogens that are handled every day by medical bacteriologists.

(a) The known laboratory infections (about 6,000 recorded in the history of microbiology) have been largely respiratory infections, spread by droplets (mostly before safety cabinets were introduced in the 1940s). Enteric infections, however, occur through swallowing of contaminated food or other material. Even the most virulent enteric pathogens are relatively safe to handle in the laboratory with simple precautions, such as not putting food or cigarettes on the laboratory bench.

(b) Strain K12 of E. coli has become adapted to artificial media during transfer for at least 30 years in the laboratory. Recent tests in England showed that after a dose in man much larger than what one would expect from a laboratory accident it disappeared from the stools within a few days. Its problems of survival are analogous to those of a delicate hothouse plant thrown out to compete with the weeds in a field.

(c) The addition of a block of foreign DNA to an organism will ordinarily decrease its adaptation to its natural habitat and hence its probability of spreading.

(d) A very large safety factor is added by the provision in the present Guidelines for biological containment. All work with mammalian DNA must be carried out only in an EK2 strain, which has a drastically impaired ability to multiply, or to transfer its plasmid, except under very special conditions provided in the laboratory.

In this connection I would question the specification, in the Guidelines, that an EK2 strain must have a survival frequency of less than 10^{-8} under natural conditions (interpreted by the committee as residual viability after 24 hours). Just as infection can be dramatically cured by a bacteriostatic antibiotic, such as chloramphenicol, as well as by a bactericidal one, such as penicillin, so the inability of an EK2 strain to multiply in the gut would be sufficient to ensure its rapid disappearance, even if it did not rapidly commit suicide. The important question, requiring extensive investigation, is not the rate of suicide of the EK2 strain, but the chance of transfer of its plasmid to a better adapted strain, before disappearance of the EK2 host.

We thus see that with a strain known to have added the gene for a potent toxin, a serious laboratory infection requires the compounding of four low probabilities. With strains from shotgun experiments we have a fifth, very low probability, already mentioned: that an apparently harmless mammalian tissue will yield a dangerous product.

The risks thus seem very much smaller than the public has been led to believe. Nevertheless, it is important to keep all the probabilities low. For example, even if a toxin-producing strain could survive only very briefly in the gut, a large enough dose might meanwhile cause disease. Hence a major benefit from the current discussion could be the requirement that those working in this area learn and use the standard techniques of medical microbiology, at least until we have acquired much more experience.

I now come to the most important point of all: the enormous difference between the danger of causing a laboratory infection and the further danger of unleashing an epidemic. In Camp Detrick, working for 25 years on the most communicable and virulent pathogens known, 423 laboratory infections were seen, most caused by respiratory pathogens. Yet, there was not a single case of secondary spread to any person outside the laboratory. Similary, in the Communicable Disease Center of the U.S. Public Health Service, 150 laboratory infections were recorded, with 1 case of transmission to a relative. Elsewhere in the world there have been about two dozen laboratory-based microepidemics recorded, each involving a few outsiders.

With enteric pathogens the danger of secondary cases is even less, for with this class of agents modern sanitation provides infinitely better control than we can provide for respiratory infection: in contrast to influenza, the appearance of a case of typhoid in a home does not lead to an epidemic. Enteric epidemics appear only with inadequate personal hygiene or sanitation, and such epidemics are always small (except when sewage freely enters the water supply).

This information is clearly pertinent to the recombinants that we are discussing. For despite widespread public apprehension about presumed biparental chimeras with totally unknown properties, these recombinants are genetically 99.9% E. coli. It is exceedingly improbable that such an organism could have a radically expanded habitat, no longer confined to the gut. It is even harder to see that the organism might be more communicable, or more virulent, than our worst enteric pathogens, which cause typhoid and dysentery. The Andromeda Strain remains science fiction.

I conclude that if by remote chance a recombinant strain should be pathogenic, and if it should cause a laboratory infection, that infection would give an early warning, thereby decreasing the chance of spread. Moreover, modern sanitation provides powerful protection against the chain or transmission required for an epidemic.

We must therefore ask whether the problem really merits deep concern by the general public, any more than the problem of how laboratories performing diagnostic work or research on known pathogens should be operated. For to produce a serious epidemic by introducing fragments of mammalian DNA into _E. coli_ would require the compounding of five low probabilities. By any reasonable judgment the risk seems very much less than that from pathogens that are being cultivated in laboratories all the time. In the United States up to 1960, 2400 laboratory infections were recorded, with 107 deaths: a small price for the millions of lives saved by investigative and diagnostic work on pathogenic bacteria.

Tumor viruses present a special problem, for any conceivable infection by a bacterium containing a tumor virus genome would lack the early warning of the toxin producers. However, all other aspects of the problem remain the same. And this loss of one protective feature is balanced by the fact that these viruses, by definition, have their own means of spread. Indeed, in general this spread is even more effective than that of bacteria, since each infected animal cell produces thousands of infectious particles. Moreover, viral DNA from a bacterium would have to reach human cells as naked DNA, and it is hard to imagine that that would be as hazardous as the same DNA in its own infectious, viral coat, already adapted by evolution for entering animal cells. Indeed, since a sensible assessment must measure risks in terms of increment over background, we must recognize that part of this background is the ingestion of the same normal mammalian tissue, as in a rare steak -- a direct source of the DNA whose transmission via _E. coli_ is feared.

While I have questioned whether the public's anxiety over potential epidemics is realistic, I do not question whether it is legitimate: there is no doubt about society's right to limit hazardous activities. However, when we ask in addition whether our increasing power to manipulate genetic material creates long-term evolutionary dangers we are moving into quite a different area, involving the concept of dangerous knowledge rather than dangerous actions. Perhaps we can clarify the issue by trying to translate into more specific terms some of the general sources of apprehension that Dr. Sinsheimer has expressed in various publications.

1. Dr. Sinsheimer questions our moral right to breach the barrier between prokaryotes and eukaryotes, since we simply cannot foresee the consequences. This argument seems to turn evolutionary principles through 180 degrees. The barriers that evolution has established between species are designed to avoid wasteful matings, i.e., matings whose products would be monstrosities, in the sense of being unable to survive, rather than monsters, in the sense of taking over. Since survival of an organism depends upon a balanced genome, evolution proceeds in small steps, no one of which will excessively unbalance the genome in one respect while improving its adaptation in another. As a result crosses between even closely related species are excluded in nature; hence it is exceedingly unlikely that artificial transfers of genes between distant species would pass the test of Darwinian fitness.

2. "The power to change the evolutionary process is as significant as cracking the atom." But atoms are not subject to extinction by Darwinian selection. Stores of nuclear weapons are likely to be more permanent than any dangerous organism that might reach the world from a laboratory working with recombinant DNA. Moreover, George Wald's statement that "a living organism is forever", though dramatic, disregards two powerful evolutionary predictions: first, that natural selection will rapidly extinguish all evolutionary departures except for the infinitesimal fraction that have improved their adaptive fitness; and second, that the recombination of genes from distant sources has an exceedingly small probability of improving fitness.

3. Power over nucleic acids, as over the atomic nucleus, "might drive us too swiftly toward some unseen chasm...We should not thrust inquiry too far beyond our perception of its consequences". I would suggest, on the contrary, that we should not thrust our limitations on research too far beyond our perception of its hazards. Some claim that scientists are arrogant in their reckless drive to explore the unknown. But considering the history of the benefits of science, and the sad history of Italy's elimination from the race by Pope Urban VIII after its head start under Galileo, perhaps it is more arrogant for a small minority of opposed scientists to believe that their convictions merit direct public appeal, with all the attendant risk of creating hysteria and contributing to a discouraging public view of science.

4. Finally, Dr. Sinsheimer suggests that this is the beginning of a genetic engineering that will ultimately lead to man. In contrast to the vagueness of the preceding proposition, this one is concrete, and one can wrestle with it. I suspect that it is what most worries Dr. Sinsheimer, and much of the audience.

In 1970 this topic received extensive discussion, and the anxiety largely subsided; but it has been reactivated by the very different question of genetic engineering in bacteria. Unfortunately this is too large a problem to consider in detail here.

Philosophical questions about the effect of science and technology on man's fate do not start with recombinant DNA: they go back to Galileo. And Max Born has made the pessimistic suggestion that the consequences of the discovery of the scientific method may conceivably hasten man's ultimate extinction -- which may conceivably be true. But the opposite may also be true. And I do not see how we can make concrete plans on the basis of such an open-ended question (and the enormous time scale that is involved).

It will not help to try to close Pandora's box. We cannot unlearn the scientific method, and if we restrict it in one place it will turn up in another. The world has only recently come to realize how large (and often unexpected) a price we are paying for various aspects of technology, and in the understandable reaction it is only too easy to take the benefits of science and technology for granted and to object to the new problems that they are raising. But in the long run it is difficult to see how we can plot a more prudent course than to try to recognize specific hazards as they arise, to seek a reasonable balance between freedom of action and protection from excessive risks, and to seek orderly and responsible methods for involving the public in matters that so deeply affect its interests.

I share Dr. Sinsheimer's concern for the future and his passionate advocacy of vigilance, and I respect his sincerity and courage. But the vigilance must be directed at specific definable applications. Vigilance concerning new knowledge that might someday be misused is, to me, a threat to freedom of inquiry, and I believe a threat to human welfare. We may conceivably be entering dangerous territory in exploring recombinant DNA; but we will surely be entering dangerous territory if we start to limit inquiry on the basis of our incapacity to foresee its consequences.

Dr. Stetten: Our next speaker is Dr. John Tooze from Heidelberg.

Dr. Tooze:

Thank you Mr. Chairman. I am going to summarize in ten minutes the efforts to formulate guidelines in a collection of twenty nation states who have chauvinistic pressures (these even reach to the extent of their wanting to speak their own languages!) But they do combine periodically in regional organizations and in this particular field three are relevant: The European Microbiology Organization, The European Science Foundation and The European Economic Community - The Common Market. Only the U.K. government decided to make an independent assessment of the question concerning the common DNA research. In the absence of any factual information, I think that most European countries felt that two assessments would be enough.

The last two speeches have demonstrated that it is difficult to reach any satisfactory conclusions at the present time. The U.K. set up a committee which made a report published last August (the Williams Committee Report). The EMBO was asked by sixteen governments to comment upon the American guidelines and the British guidelines based on assessment (in the absence of very much data) of conjectural hazards and make some recommendations as to what all the other countries might do. This report will be published in NASUM and I would like to summarize briefly the major points.

The first is that within the United Kingdom (since the British government adopted this report) genetic engineering research is to be continued and promoted. There is no talk of a general prohibition or a "let's wait around while we develop a new host other than E. coli". An adjective is used continually to describe the hazard - it is "conjecture", "conjectural", and "conjectured". But, of course, this research is to continue under certain precautionary conditions. These involve physical containment, biological containment and certain implementary measures. It is quite clear that the British government gives greater emphasis to physical containment than to biological containment. It, like the American guidelines, set out four categories (1,2,3,4). The British categories 1,2, and 3 are noticeably more stringent than U.S. categories P1, P2, P3. C4 and P4 are roughly the same - you have reached more or less the peak of what one can imagine of physical containment.

The British government in the report it accepted did not define in any great extent the disablement of bacteria or bacterial strains, host and vectors, for this research. It accepted that the host and vectors should indeed be disarmed, and disabled, but it refrained from giving any precise definition, any number 10^{-8}, 10^{-12}, 10^{-11}, 10^{-6}, 10^{-4}, and so on. But then, for reduction and survival, the argument was that it is better to decide each case one by one in the context of the experiment by experimental protocol. There are some legislative differences - very important ones.

First of all, in Britain all experiments of this sort are legally reportable to a central committee which was established as a consequence of the Williams Report under an act called the "Health and Safety At Work Act", and this applies to all laboratories, be they industrial, military, etc. Secondly, the GMAG (Genetic Manipulation Advisory Group), which is to regulate things in Britain, is independent of all funding agencies. It contains representatives of the Trade Union Council, two unions representing scientific workers, of the public (one of the choices is John Maddox, former Editor of *Nature*, to represent the public), medical microbiologists, and scientists able to judge this work (but none of them actually engaged upon it as far as I know - people like Mark Richmond, Bob Harries, Brooksby, etc.). This machinery has been set up and the Committee has met twice. A decision on one experiment has been made.

Faced with this situation, the EMBO Committee decided that in each European country it would be politically inevitable to set up some national advisory group. It should be a central group but, depending upon the country, this could be either governmental, research council or academic. There are subtle differences and they vary from nation to nation. The EMBO report said that all experiments in any of these countries should be reported to the National Advisory Group regardless of the source of funds for the laboratories involved. Most importantly, they recognized that for small countries (a country with a population of 4 or 5 million) no bigger than one of your small states, it is very difficult to operate the British procedure because it relies on a GMAG which makes independent assessments of each and every experiment's physical and biological containment. It would be unrealistic to recommend this in some European countries.

Therefore, the Committee proposed that within Europe either the NIH Guidelines or the Williams Guidelines could be applied and used. This means that a country such as Norway (4 million souls and probably only 40 molecular biologists) can use a P3 facility or a European one, can use NIH certified material and can do the experiment as safely as it could be done in the United States. Of course, the Norwegians could barely set about establishing a GMAG. The EMBO Committee also recommended something that tends to be forgotten - that the collection of information about the viability of recombinant DNAs, of their hosts and of the vectors in the wild, and also the results of risk-testing experiments (designed to maximize the risk, but done in a very high containment so as to see whether there is any risk), should be promoted.

One such experiment, which is a version of the one that is going on in the United States, which involves putting polyoma plasmadina colina in a mouse and seeing what happens will be done in P4, as long as we get the authority of the British committee in the next 4 or 5 weeks. Most of the European countries have indeed set up national committees. The national committees have taken note of this report of the NIH and of the Williams Guidelines and most of them are now drafting guidelines for their own local use. That means taking various measures from the existing procedures that have been published and adapting them to the local, political, social and economic structures. On top of this the European regional organizations, notably the European Economic Community (which is a powerful economic body), are likely to issue a directive saying that all research in industry in the nine countries of the EED shall be reportable to the national committees. This is directly aimed at industrial research.

Countries like The Netherlands are envisaging specific legislation to make reporting compulsory legislation specific for common DNA research. The Health and Safety in Work Act in Britain is a very broad act - it just says the employer must not harm his employee or the environment - so anything can be put into it. The European Science Foundation has likewise asked for some sort of harmonization. It is quite clear that when individual decisions are made, there will be divergences, this is inescapable. In order to reveal this situation within the European community (the gross community - not the EEC) provisional authority has been made for quarterly meetings of one technical representative of each national advisory group. I think the function of that committee, if it is to work, is to say, "Dr. Jones proposes experiments with this lamba, Dr. Smith proposes another experiment with that lamba. We graded this, you graded that. Are they roughly comparable or not? If not, who has made the mistake?" That is the way we are going to proceed.

Sir Robert Williams, who was the chairman of the Committee that wrote the Williams Report, is on record as saying he believes it would be wrong at the present time to try and reconcile in every detail the NIH Guidelines and the British proposals, and therefore follow all the other combinations that will emerge in Europe. I personally believe this is a very correct position for, at the present time, we are totally ignorant of the properties of recombinant DNA molecules and most of the things we are going to use to make them. Likewise it seems to me that if we are all dragooned into following one path and that is the wrong path, then we will all end up in the wrong place. On the other hand, if we all follow roughly parallel and similar paths, then we might more readily obtain more relevant information so that the long-term aim of dismantling the bureaucracy, if that is the way the thing goes, can be achieved. If it means establishing more bureaucracy, then that will be the consequence.

I think the only sort of philosophical point I would choose to make is the following: that we are attempting to regulate the freedom of scientific inquiry. As a start there are two options, either it can regulate the acquisition of knowledge, or it can regulate the application of knowledge. In my personal opinion, it is not only easier to regulate the application, but in the long run, it is infinitely more desirable to regulate the application than to try and regulate the acquisition of knowledge. Thank you.

Dr. Stetten: Our next speaker is Dr. A. Bayev of the Soviet Union.

Dr. Bayev:

Thank you. Recombinant DNA research in the U.S.S.R. had hardly begun at the time of the Asilomar Conference. Since that time things have changed and at least five or six laboratories in the U.S.S.R. are involved in recombinant DNA research. This is not very much in comparison to the United States, but it is large enough to cause concern over the potential biohazards of recombinant DNA. Now we will begin to prepare guidelines on recombinant DNA and encounter many delays since I think some experience in recombinant DNA research must be accumulated before the guidelines can be devised in our country. It is a new field for our scientists.

It is my personal opinion that we should take the NIH document as a basis for our guidelines and there are some reasons for that.

First of all, we greatly appreciate the work of the American scientists, particularly Drs. Berg, Stetten, Sinsheimer, Szybalski, Baltimore and many others. We also appreciate the reports done by NIH and the National Academy of Sciences. I am certain that work of this nature could not be done by us or anybody else in such a short period. But, the second reason is more important in my opinion. The environment is indivisible; it is common for all countries; for all human beings. It would be contradictory to common sense to follow different paths and use different means for the protection of the environment and the people. It is the duty of scientists to formulate the right policy in this field at the very beginning. By doing so we may avoid otherwise inevitable mistakes and prevent the potential deterioration of the environment.

Research in recombinant DNA should be done, but under proper conditions without any danger to human communities. I think that the international cooperation of scientists is very important at a time when there are no serious political and social obstacles. Thank you.

Dr. Stetten: Thank you Dr. Bayev. Our next speaker from the National Cancer Institute is Dr. Maxine Singer.

Dr. Singer:
I would like to talk about the National Environmental Policy Act of 1969 and the ways in which it impinges on the recombinant DNA issue. One of the stated purposes of this Act is "to promote efforts which will prevent or eliminate damage to the environment and the biosphere and stimulate the health and welfare of man". The Act charges the Federal Government to use all practicable means, consistent with other essential considerations of national policy, to fulfill the responsibilities of each generation as trustee of the environment for succeeding generations and to assure for all Americans safe, healthful, productive, and aesthetically and culturally pleasing surroundings. Thus, the Act covers an extremely broad range of activities, from construction to health measures to the preservation of natural and historical sites.

The primary mechanism established by the Act to attain its objectives is the requirement that all agencies of the

Federal Government consider environmental consequences in the development and carrying out of all federal policies and programs. Most specifically, all proposals for legislation and all other major federal actions significantly affecting the quality of the human environment require preparation of a detailed statement called an Environmental Impact Statement, by the responsible official. The Statement is to include a discussion of the environmental impact of the proposed action, including unavoidable, adverse impacts and alternatives to the proposed action.

Consistent with the wording of the Act, the current procedure has the agency in question preparing a draft statement, publishing and circulating that draft statement for comment by other interested federal agencies and organizations and individuals, revision of the draft in response to the comments, and finally the filing of the statement to the Council on Environmental Quality, the Council that was set up by the Act and exists within the Executive Office of the President. It is the responsibility of the Council to apprise the programs of the Federal Government in the light of the aims set forth in the Act.

The Act itself, is a brief 2-1/2 pages long, in marked contrast to the environmental impact statements that it generates, some of which are thousand of pages long. Reading the Act gives no clue to the magnitude of what has grown from it. Within each federal department a special bureaucracy now exists to monitor compliance for the Act on the part of the several divisions of the department. Within HEW, for example, the Department of Health, Education, and Welfare (which is the parent of the NIH), lengthy and detailed rules have been laid down to govern the preparation of statements and to attempt to be sure that the contents of the statements are consistent with the policies of the department. Reading the brief Act, one would also not imagine the amount of work the Act has generated for the federal judiciary. Concerned individuals and private organizations have questioned in the courts virtually every aspect of environmental impact statements. Some legal commentators have suggested that the lack of precision in the Act, about just what the Act is intended to achieve, is responsible for the proliferation of legal cases. It has, nevertheless, become the chief and most successful tool of the environmentalists in this country. Some lawsuits have questioned the propriety and legality of an agency's rules governing the preparation of the statements. Some suits have challenged an agency's decision when an impact statement was not needed in a given instance and, for example, judicial decisions have gone so far as to say that the Act must be construed to include protection of that vague but fashionable attribute, the quality of life.

These were the first questions with which the NIH had to deal in considering the relevance of the Act to recombinant DNA. In what sense is the recombinant DNA situation a federal action? And, finally, in what way will the activities of NIH, in regard to recombinant DNA, significantly affect the quality of the human environment? Obviously there are several possible answers to each of these questions.

The NIH adopted the following views: first, that the publication of guidelines for the conduct of recombinant DNA research constitutes a major federal action in part because of the widespread interest and debate on the issue; and second, that since one cannot state with any certainty whether the research and guidelines will significantly affect the quality of the human environment or not, the public interest and intent of the Act will be best served by the preparation of a statement and the consequent opportunities for public education and discussion.

The discussions leading to the decision to prepare an environmental impact statement at the NIH were coincident with the development of the Guidelines themselves and it became evident that a serious problem in timing existed. According to the National Environmental Policy Act, the publication and circulation of the impact statements should proceed prior to the action itself, - in this case the publication of the Guidelines. The preparation of an impact statement is a large and usually lengthy undertaking and although the discussions leading to the development of the Guidelines involve much of the type of analysis of possible hazards, possible benefits and alteratives that are required in the preparation of the statement, there was still a great deal of work to be done to prepare an adequate statement and to have it approved for publication by the HEW hierarchy. The Director of NIH decided to proceed with the publication of the Guidelines last June, although the draft Environmental Impact Statement was not ready and was not published until early September. As the Director recognized in the decision statement accompanying the Guidelines, his action was not in compliance with the Act. Nevertheless, the Guidelines were published because the Director believed that they afforded a greater degree of scrutiny and protection than the purely voluntary, and less specific, Asilomar Guideline that was then in effect. He believed, and still does, that this action, inconsistent with the law though it was, was the best course and most in keeping with the intent of the law, because the NIH Guidelines represented a higher degree than the Asilomar Guidelines and the public interest was therefore best served by prompt publication.

This position has been criticized by some, but it is my belief that this criticism stems from the false impression that prior to the publication of the Guidelines in June there was a moratorium on all recombinant DNA experiments in the United States. There has never been such a moratorium in this country. Many environmental impact statements are prepared for federal agencies by independent organizations on contract to the agency; as a matter of fact a whole industry has been built up in this regard. The draft Impact Statement on recombinant DNA guidelines was prepared by the NIH staff, both scientific and administrative, with the help of many investigators across the country. It was not a long document by environmental impact statement standards. It attempts to be impartial and non-judgmental except in the section describing the action chosen by the NIH, - in this case, the Guidelines that were published.

The Impact Statement describes the objective of the action, that is, the protection of laboratory workers, the general public, and the environment from infection by possibly hazardous agents that may result from recombinant DNA research. It has a brief description of the nature of the experiments, written assuming a naive reader, and it summarizes the events that led to the development of the Guidelines. After this, the Impact Statement goes on to discuss the issues raised by recombinant DNA research, possible hazardous situations, and how theoretically they might arise, and expected benefits. The conjectural nature of the possible hazards and many of the potential benefits is stressed and no attempt at advocacy is made.

The Impact Statement then summarizes the proposed action (the Guidelines), and attempts to explain the factors that went into the assessment of possible hazards that is implicit in the rankings that are found in the Guidelines. Alternatives to the recommendations in the Guidelines are discussed - no action at all by the NIH, or a prohibition on funding of all experiments involving recombinant DNA, or the development of a different set of guidelines, for example, exclusive reliance on P4 conditions or guidelines not permitting the use of *E. coli*. Alternative mechanisms for enforcement and for implementation are also discussed, as is the alternative of a general federal regulation for all such research.

In the final section, an attempt is made to evaluate the impact of the Guidelines on the safety of laboratory workers and the environment and other humans in that environment, on the cost of doing experiments and on the efficiency of getting work done. The impact of the experiments themselves is discussed, both in terms of potential hazards and beneficial impacts. The Environmental Impact Statement on recombinant DNA guidelines will not make pleasurable reading for those who adhere strongly to one extreme or the other of the current controversy. It has made such readers angry and has disturbed their personal environment. For those who are willing to recognize what the Impact Statement does - that in the absence of all the needed facts we must still make reasonable and rational judments, that total security from risk is probably equivalent to total stagnation, but that laissez-faire is irresponsible - a reading of the Impact Statement may help establish an analytical framework for serious consideration of the problems that trouble all of us.

Dr. Stetten: Our next speaker is Dr. Paul Berg.

Dr. Berg:
I wish that I had some new information to bring to you and to this issue as some of the former speakers have already done. I would really like to take just a few minutes to recapitulate on some thoughts that have come to me as I have listened to the presentations today and also some thoughts that I have agonized over through the past few years.

There is always a problem of matching Dr. Sinsheimer's eloquence and the way he states his views. I believe the sincerity of his views but I also believe that his eloquence in speaking on this issue should not be taken as indicative of a monopoly on wisdom or on clairvoyance of the outcome of recombinant DNA research. It is his point of view which he has pushed vigorously in many different arenas.

Dr. Sinsheimer in opening, read a definition of a dialogue and I have tried to think whether the past four years (and it has been four years in which this issue has been widely discussed) meets the criteria that he posed for such a dialogue. True enough, the issue was raised by scientists informally in the beginning of June 1972. Only a limited number of people were involved in that discussion but, nevertheless, the discussion widened as time went on. We have now had four years in which this problem has been reviewed, discussed, debated, written about and filmed. I am sure that I have left out other adjectives describing the dialogue, but it can hardly be identified as something that has been sequestered behind closed doors and confined to the scientific community alone. We have had citizens groups, lay groups, and educated non-scientists (both here and abroad) who have examined the issues that Dr. Sinsheimer thinks need to be reviewed. The debate has been open to the public and the press. At no point have I ever heard of anybody being uninformed or prevented from being informed on every nuance of the discussion or issue. I believe it is the most widely debated and publicized scientific issue of any I have encountered.

Dr. Sinsheimer's views on recombinant DNA are clearly a minority. It certainly has had its hearing in many different circumstances and the overwhelming mass of opinion and judgment on the part of other scientists is that the risks to scientists, their families, the public, and the environment are small small enough to permit us to proceed under the conditions of the NIH Guidelines. It is impossible to say that the risks are zero, and I don't think any of us can expect any human endeavor to proceed with a zero risk. I am encouraged by several developments during the last four years and one of them is the extent to which independent reviewers have arrived at pretty much the same conclusion.

Five months ago the Cambridge City Council appointed a committee to examine the issue of the advisability of proceeding with work that requires P3 type facilities and whether to permit Harvard University to build a P3 facility to permit individuals to work under those conditions. The membership of the committee consisted of primarily non-scientists - certainly not committed or in any way involved in recombinant DNA research. (I don't know the roster, but I am certain that there were teachers, housewives, and workers from throughout the Cambridge community.) After five months of intensive study (and I am told that they educated themselves remarkably) they came to the conclusion that P3 type research should and could proceed in Cambridge and that the University should be permitted to build their P3 facility. I find that kind of independent review and the seriousness with which they took on their task and prepared their report as refreshing. I think that it belies Dr.

Sinsheimer's charge that it is the scientists' self-interest or ego gratification that has led them to proceed without concern for anyone else in the world.

I think I have also learned that it is virtually impossible to expect any unanimous agreement. I think we certainly want to have informed consent, but I think it is impossible to expect that we will have unanimous consent. I am persuaded on the basis of my study (and what people who are more expert than I have educated me to) that the Guidelines are sufficient, and perhaps even more stringent than is necessary; but nevertheless sufficient to meet the perceived risks. I am prepared to go ahead and work under those conditions as I believe most scientists in this country and throughout the world have also agreed.

Dr. Stetten: Thank you members of the panel.

INTER-PANEL QUESTIONS

Dr. Stetten:

I am going to take the Chairman's prerogative and ask the first question myself. I am going to address it to Dr. Sinsheimer.

It has appeared to me that there is frequently a morbid discrepancy between what I perceive as the magnitude of a hazard and the magnitude of society's anxiety toward that hazard. Let me give two examples. We have an epidemic of an unknown disease in Philadelphia which claimed roughly a score of lives and made about ten times that number of people ill. The nation, the press, and the media reacted with what I felt to be a very real wave of anxiety. This is the same nation which we have been totally unable to scare with respect to cigarettes despite real efforts by the Surgeon General. 50,000 to 75,000 people will die this year as either direct or indirect results of smoking cigarettes and we can do nothing about it.

We have enormous anxiety. I see it every summer at Woods Hole concerning the hazards of being attacked by a shark while swimming. We have similar anxieties concerning the hazards of being bitten by a poisonous snake I am told that both of these events occur with extreme rarity in this country a country which will sacrifice roughly 50,000 lives this year to the automobile about which no one has any anxiety whatsoever. I would like to ask Dr. Sinsheimer whether in the first place it is a general truth that there is little relationship between the magnitude of anxiety and the magnitude of hazard; and if this is true, does he have any explanation for it? Why are we such bad judges?

Dr. Sinsheimer:

Of course, you are correct that the magnitude of societal anxieties is affected by all kinds of cultural hang-ups and what not. I am certainly not qualified to tell you why all the campaigns with respect to smoking have no effect - maybe somebody else here may be better qualified...I guess that I am not quite clear on what you are trying to tell me. Are you trying to tell me that the fact that people are concerned about recombinant DNA is something I should not take seriously?

Dr. Stetten:

No. What I am trying to say is the magnitude of the anxiety may not be any measure at all of the magnitude of the hazard.

Dr. Sinsheimer:

True, but if I were a snake, I'd worry about the public's fear of snakes!

Dr. Stetten:

Our Presiding Officer, Dr. Schultz, would like to ask a question.

Dr. Schultz:

Mr. Chairman, I am concerned about the magnitude of the anxiety of those scientists in this field (some of whom have already spoken with me this week) who have stopped work on research they would like to do because of the stringency of the requirements for P4. Their institutions are incapable of providing adequate funds for the construction of P4 facilities. Are the P4 facilities a little bit more than we need since you and your committee have reacted to these anxieties to provide extremely stringent requirements? What is the cost of a P4 facility versus a P3?

Dr. Stetten:

I can certainly give you rough figures. The moving van which has been converted into a P4 facility (which is the only one that I know of which could be accessible in about six weeks) cost a third of a million dollars eight or nine years ago. Today it would cost at least a half a million dollars to replace the van. We are contemplating redoing some existing buildings at Fort Detrick which contain about 20,000 gross square feet of space. The buildings exist and can be converted, we are told, into P4 facilities at a cost of about three million dollars. I am also told that a given working laboratory, depending upon the air-flow arrangements (air-conditioning and other things), can be converted into a P3 laboratory for approximately twentyfive thousand dollars. I have not priced this out, but I am told this is so.

Dr. Sinsheimer, do you have a question?

Dr. Sinsheimer:
I have a couple of comments rather than questions. First of all, I really would like to react to what Dr. Berg said. I don't think he quite meant it. I have not said that it was the scientists' self-interest or ego gratification that has been involved. It is simply that scientists, because they are scientists, have different value sets than other groups of people. That is perfectly understandable and I do think that their influence in this dialogue so far has certainly been very disproportionate to their numbers of the population. You may say that it should be disproportionate and we can discuss that. But I just want to make it clear that I am not attributing this to self-interest or ego gratification - that is not what I am trying to say.

Secondly, I would like to make a brief comment on what Dr. Davis said because it seems to me that in my statement I made some reference to the fact that I felt we did not have general theorems which permitted us to predict the consequences of recombinant DNA experiments. In those terms, it seems to me what Dr. Davis is saying is yes, we do have a general theorem - we have Darwin's theory of natural selection and that takes care of the whole matter. I am not sure that Darwin would agree. Maybe that should be the Darwin-Davis theorem. I think you are being too modest Dr. Davis; I think you are pronouncing a whole new theorem which is the theorem not that organisms are well adapted to their environment, but that they are perfectly adapted to their environment and we could not possibly make a better one. You seem to imply that everything that has been done could be done, and I don't believe that. I don't see what the evidence is for it. I'll stop right there.

Dr. Stetten: Dr. Singer do you have a question?

Dr. Singer:

I have a couple of comments which lead to a question. I wanted to address myself to Dr. Sinsheimer's criticisms of the process that has gone on over the past few years leading to the establishment of a Federal Government policy governing recombinant DNA work. I think that Bob's description of an ideal process in which a country establishes certain general principles for broad policy and then works out the technical details is a nice one, but it does not work. For that reason in general, policies are made the other way around as indeed this one was made. Doing it by establishing general principles in a democratic society usually means that you get hung up on getting agreement on general principles. In a pragmatic way it is much simpler to get some sort of consensus on small technical issues one by one - that is backed in the history of the way the government of this country works on most things. I think it recognizes what is realistic and how one can achieve most promptly an appropriate action. In fact, in this process, the matter has been taken up within the Congress. There have been two hearings held by the Senate Subcommittee on Health (one in May of 1975; one in September of 1976) which dealt with the issue, and Senator Kennedy, the Chairman of that Committee, addressed himself to what was going on within the NIH and within the scientific community. By the fact that he did not take any further action on this he appeared to be saying that things were moving in at least a reasonable and acceptable way.

Most recently there has been growing discussion of this issue in local government situations. The Cambridge City Council is but one example. Since Dr. Sinsheimer, several others and I, myself, have taken part in hearings of various kinds in different parts of the country, I would like him to comment on how he views this particular new situation of a great deal of local action and the attempt by local government to deal very specifically with these issues.

Dr. Sinsheimer:

Certainly. I would like to respond to the earlier part of your statement as well. I am sure you are right but normally we do probably proceed from the particular to the general, rather than the other way around, but I think that caused a greal deal of difficulty. I might point out - and I am not saying that it's necessarily a good example - but Congress wrestled for six years until last session it finally passed the Toxic Substances Bill which is to regulate the introduction of chemicals into the environment, and in the course of that a great deal of policy was developed. I am not saying it should take that policy and apply it directly to recombinant DNA since each case is different. But I do think you have to develop some kind of policies. Otherwise there are policies almost implicit in the technical details which may not have been well thought through.

To answer the matter of this discussion in localities, as Dr. Singer knows, I have not been involved directly at all in the discussions in Cambridge. There are plenty of people in Cambridge able to present either side of the issue without needing me. I do think however that it is really impractical to handle this kind of an issue on a local basis. It just doesn't make any sense. You are not significantly safer if the research is banned in Cambridge, but goes on across the Charles River in Boston. Indeed, attempting to handle it on a local basis frankly biases the issue in a direction which I would argue is a poor tactic if nothing else. It biases the issue because people in it will always raise the argument whenever it is mentioned that if you ban it here, you will just lose our scientists (and they are correct) and this prejudices the outcome.

I would, for that reason, have to take a slight exception to what Dr. Berg said with regard to the result of the Cambridge Committee. I don't mean that they didn't do a good job, but I am sure that one factor weighing on them is just the one I mentioned that if they were to ban the research in Cambridge, people would leave Harvard and MIT and they really wouldn't have accomplished very much.

Dr. Stetten: Thank you. Dr. Davis, would you care to ask a question?

Dr. Davis:

Yes, I would like to make a brief comment in reply to Dr. Sinsheimer's comment on mine. If I had offered a series of deterministic propositions I would say that his criticism would be absolutely correct; but since I believe that everything I said was framed in terms of probabilities, I don't think I exaggerated the implications of Darwinism. That is my view.

I would like to ask a question of Dr. Singer about the Environmental Impact Statement. Studying that statement and the preceding guidelines, I could not help feeling that there was a distinct difference in tone. They both tried to be neutral and address themselves to pros and cons of a whole variety of issues, but I felt there was considerable weighting in the direction of the hazards, and much less in terms of the reasons to think they were not so large in the second statement - the Environmental Impact Statement. I know nothing concerning the history of how they were developed. Could you offer any explanation for this difference, or whether or not you perceive that there is such a difference?

Dr. Singer:
I would agree that there is a difference. I think it is in the nature of the two documents. The Guidelines specify a policy which was arrived at on the basis of making certain judgments. The facts and the considerations that went into those judgments are what is laid out in the Impact Statement, which is designed, in order to be in compliance with the law, to be as non-judgmental and as straightforward as possible. Perhaps your comment is put in some perspective by my telling you that some of the comments that have been received on the Impact Statement say precisely the same thing - except reversing the words "benefits" and "hazards". So, it depends on who you are. Huntza tells the story about knowing you are in the right place going through a narrow strait when you can hear the crashing of the water equally well from both sides. I conclude from this that the Impact Statement is not too far from the mark.

Dr. Stetten: Dr. Berg do you have a question?

Dr. Berg:
I would like to make just one other comment. Not only do we often meet on podium or table like this before meetings, but we also correspond with each other on occasion about points of view and things that arise concerning this issue. Both by correspondence and at this meeting I have had the opportunity to explore with Dr. Sinsheimer (at least in more detail) what he feels he has been advocating and what I believe is a misconception about what the audience throughout the country and the world thinks he is saying. If I can at least say what I think is the distillation of our discussion, and then allow Dr. Sinsheimer to amplify it - it is that whereas local communities and other groups who take him as their champion have assumed that Dr. Sinsheimer is advocating stopping P3 type research in the various locales where he visits. Therefore, they have almost taken action along that advice. He tells me that that has not been his advice to these groups. I am not saying this very well. The point is that whenever I have gone to any committee that has had a hearing, they always confront me with the fact that Bob Sinsheimer has advocated that we not permit P3 type research

to go on - he tells me that that is not the kind of advice that he has been giving and that he has a much broader view of the problem and advocates a national or international cessation of such work and to relegate it to special facilities where its supervision could be guaranteed more efficiently. Dr. Sinsheimer, would you want to comment on that further?

Dr. Sinsheimer:

Certainly. Basically, as I have said before, I would prefer to see all the work which is permitted in P3 facilities under the Guidelines, be done under P4 type containment. But, I do not believe it is practical to attempt to achieve that by local action. When I have been asked (and I'd like to make another point that I don't go around asking to be heard at these "seances" - they invite me and then ask my opinion), the advice that I have generally given is that yes, I think these experiments are hazardous;that I don't think it can be handled on a local basis; that if the responsible body (Board of Trustees, City Council, etc.) shares my opinion they should try to make that opinion known at the national level to the Secretary of HEW or to their senators and exert their influence to have this issue reopened and rediscussed at a national level. Ultimately, one would have to take it up at an international level because the hazard is a global one. As Dr Bayev mentioned - you are no safer if some new agent is developed in Mexico than if it is developed in Boston. But, the international problem is one we cannot deal with if we cannot first achieve some position nationally.

Dr. Berg:

The point I want to reply to is that since we have been unable to achieve international agreement on much more serious issues and since the likelihood of achieving international agreement on a set of containment requirements which you feel are adequate for this work is so small, would you advocate following the NIH Guidelines as they are now written until we can achieve this almost unattainable goal?

Dr. Sinsheimer:

I would like to see a great many of the experiments permitted in P2 and P3 facilities under the Guidelines be done in P4 facilities. I think that if the United States had such a requirement that it would exert considerable influence on other countries. It would make scientists in other countries have to answer to their governments as to the reasons the United States adopted such strong requirements.

Dr. Stetten: Dr. Bayev, do you have a question to ask the other panelists?

Dr. Bayev:

I am not involved in ad hoc American discussion in recombinant DNA, but I can state that my personal opinion is very moderate. We must continue investigations of recombinant DNA, but we must do it under proper conditions. I should like to remind you that at the very beginning of the microbiological era there were opinions that laboratories would be very dangerous for people because they would accumulate bacteria and pathogenic agents. Now we see that there was no real danger and we have no dangerous consequences. I think that after setting proper measures, we will be able to work with recombinant DNA without danger. It is only a question of time.

WRITTEN QUESTIONS

Question 1:

Is there any evidence that natural mechanisms and vectors are now in operation and are currently causing an exchange of DNA segments between human genomes and the genomes of other eukaryotes and prokaryote life forms? Is any research now being funded which examines this problem?

Dr. Berg:

As far as I know there is no hard data bearing on this question and it was originally intended that one of the major responsibilities of the Recombinant DNA Molecule Program Advisory Committee would be to identify and promote research that could help assess any of the potential risks of recombinant DNA research and certainly this area of research was one that was in that category. There is only one experiment that I know of - that is an attempt by Joe Sambrook to assess whether a eukaryote viral DNAse element could be taken up by bacteria and maintained through many generations. From our conversation earlier this week I understand that when E. coli was exposed to polyoma DNA under conditions where they are known to take up foreign DNA as by exposure to calcium, they do acquire sequences homologous to polyoma DNA. In what state these sequences are maintained through many, many generations of growth has not yet been determined. These clones, according to Joe, have been frozen away for some future time when perhaps facilities are available to do the analysis more safely. At the moment I don't know that anybody is grossly concerned with polyoma but at least in this particular experiment the implication is that polyoma DNA can be taken up and it can be maintained. It is not clear whether it is integrated into the chromosome, maintained as an episomal element or what. That is the only indication I know of concerning the maintenance of a eukaryote DNA in a prokaryote for at least the number of generations that he has carried it.

Dr. Davis:

I am not familiar with any experimental work on this but such things have gone on in the past. I could only mention that there is work published by Todaro pointing out close relationships between some of the sequences in some tumor viruses (I think they are found in cats and certain primates) from which he has drawn conclusions about the distribution of DNA really crossing rather broadly species lines through the mediation of viruses.

Dr. Sinsheimer:

It seems to me that Todaro postulated that some time several million years ago the monkey virus passed into the cat, or vice versa, but events that take place once in ten million years are really irrelevant. If people really believe that there is a lot of exchange going on between mammalian DNA and bacteria, then a lot of things that people are doing are being done the hard way. Instead of making a bacterium which has a gene transform, we should be able to pick it up out of the sewage!

Dr. Stetten:

May I comment on that? There is a recent publication of which some of you are certainly aware in which a gram-negative rod was tentatively identified as Escherichia coli. Cohen and Strampp of Princeton Laboratories, Inc., found a protein in the medium in which an E. coli was grown, a material which was immunologically and pharmacologically indistinguishable from human chorionic gonadotrophin, the organism having been isolated from the urinary bladder of a patient with colonic cancer. This was a second experiment of this sort; they made reference to an earlier comparable isolation - in this case from a gram positive coccus.

There is work going on at NIH in the laboratory of Leonard Kohn, in which a polypeptide sequence has been isolated from the enterotoxin fraction of the cholera vibriol organism which appears to be remarkably similar and to share pharmacologic action with human thyroid-stimulating hormones and I have been told of a similar isolation from

Clostridium tetani. Clinically, it has long been known that patients suffering from infections of the tetanus organism frequently exhibit something which clinically resembles a thyroid storm in which there is an abrupt and very dramatic increase in pulse rate, respiratory rate and body temperature. This clinical observation was what suggested looking in this organism for thyrotropic activity which indeed apparently has been recovered. I don't know what these experiments mean, but they are certainly compatible with the notion that organisms that infect the human may pick up some information from their host (as indeed we all do when we go visiting).

Dr. Berg:

Just to highlight what I think is the "rootal" portion of the differences that keep recurring. There is a rather sharp difference between Dr. Sinsheimer's comments on Todaro's work and what would be my perception from the fact that what Todaro reported, if it is indeed evidence of a transfer in nature millions of years ago, is clearly a very rare kind of event. I think he drew the conclusion that if that were really happening all the time we would be working the hard way - we ought to get in a little sewage. But, he is referring to the production of a recombinant that survived and I am talking about the production of recombinants that will not survive. Ones that survive I maintain will be rare events.

Question 2:

What work is being done to evaluate the possible risk of escape of vectors carrying foreign DNA? Why isn't more of this being done, or is it, and by whom?

Dr. Tooze:

I recall some remarks earlier this week when people complained about the time it took for vectors to be developed for experimental use. It seems to me that some of the people who wish to use this technology should devote slightly more time to the question of providing their own hosts and vectors - then some of the bottlenecks would be removed. I do believe that it is a very fashionable pastime doing the very elementary experiments required to test some of the ecological properties of E. coli and sewage. That of course is the reason they are not being done everybody chooses to avoid doing them until they are obliged to do them, in my opinion.

Dr. Singer:

There is an experiment which is proceeding and being done by Wally Rowe and Mal Martin of the NIH to answer the question as to what happens if you are doing a polyoma DNA segment to a coli vector system and insert that recombinant into coli and then infect germfree mice with that coli and look for a positive sign of infection by a polyoma. A lot of the preliminary work for that experiment has been done. They are ready to actually put the recorded DNA into the E. coli. That is an experiment which they will have to do under P4 conditions and they are currently awaiting the certification of the trailer Dr. Stetten alluded to before they can proceed.

Question 3:

In 1969-70, following the isolation of the lac operon by Shapiro and Beckwith, there was much discussion as to whether scientists should be doing research on the development of techniques that might allow genetic engineering of humans themselves. Since the advent of methods for making recombinant DNA molecules, discussion has centered almost exclusively on how best to perform this type of research to prevent potentially dangerous molecules from escaping from the laboratory. What happened to the discussion as to whether genetic engineering research should be done at all? Are there areas of science such as genetic engineering that should not be investigated because society is not yet prepared to cope with the moral problems that are likely to result from such research?

Dr. Berg:

Well, I thought Dr. Tooze's comments earlier were related to this question and I share his views so stringently that I cannot add anything to them.

Dr. Davis:

Well, I think that the last sentence really is not correct because the aim of genetic engineering in man, the medical aim of genetic therapy, is to replace single well-defined chemically identified genes that are defective. What people are afraid of is that once we can do that we will be able to make available to tyrants, governments, and all kinds of evil people the power to do similar things with genes affecting personality. This was pointed out in detailed articles back in 1970 and it may have been one of the reasons why the discussion subsided. There is a vast difference between replacing a gene that you can define, that makes a protein that you can define, and replacing some unknown number of genes of unknown structure, doing unknown things to the development of the brain, apart from the fact that once an infant is born its brain is mostly developed and what you do with the genes is not going to make that much difference to the learning diagram after all.

Dr. Sinsheimer:
I would just like to add one epilogue to what Dr. Tooze said. He said, quite correctly, you can have a choice of trying to inhibit the acquisition of new knowledge or trying to inhibit or control the application of the new knowledge when you get it. But the point that, I think, is overlooked is that if you choose the latter (which most scientists probably would) there is a cost. Society will have to develop ways of controlling the uses of any new knowledge that is acquired and that may require significant societal change and that is an issue which needs to be discussed more broadly.

Dr. Tooze:
I would like to associate myself entirely with the last statement of Dr. Sinsheimer. It does seem to me that it ought to be better to regulate applications, but to have freedom of acquisition of knowledge. It seems to me that the policy discussion (that scientists should be discussing with the politicians who represent the people and public involvement), should cover the sorts of societal changes that will be needed to cope with the new technological possibilities. That is what you should be discussing - not whether Cambridge will let you do an experiment in Harvard (which is irrelevant, because if you can't do it in Harvard, you can do it in Cambridge, England)!

Dr. Sinsheimer:
Well, if we're not discussing it, we are actually working out the details of how it eventually will come to that. I think that is what Maxine was saying - we will deal with it in a practical way because in the debating of how to deal with this issue we are really working out the format and some of the principles that will be applied to future problems.

Question 4:

Do you think it is of much use to restrict the research activities of NIH grantees while other workers, especially those from industry and outside the United States, have no such restriction?

Dr. Stetten:

I can respond to part of that question. The NIH has been named by the White House as the lead agency in seeking cooperation among the several agencies of government. I think this has been secured. We have had an inter-agency committee established in which all of the concerned agencies are represented and there is substantial agreement among all governmental agencies that are likely to do such research that they will subscribe to the NIH Guidelines.

Far more difficult is the problem of which, if any, agency will be called upon to regulate. The NIH takes pride in not being a regulatory agency. Among the regulatory agencies that are closest to this problem are OSHA (Occupational Safety and Health Administration), EPA (Environmental Protection Agency), and CDC (Center for Disease Control). We have had meetings with the representatives of the General Counsels' offices of these agencies and at the last three meetings I think it was quite generally agreed that no one agency seems to have the authorization to regulate all phases of this work. There is a flaw or an imperfection in each of these authorizations from the point of view of this problem.

These legal experts have been asked to develop legislative drafts for submission to the Secretary of HEW to achieve the necessary degree of control which could be applied to and could be accepted by industry and possibly other agencies beyond the scope of NIH. NIH's sole sanction is to withhold funds.

Dr. Stetten:

The panel will now entertain questions from the floor.

FREE QUESTIONS

Decker, Columbus, Ohio:

I would like to make a comment particularly relevant to the title of this panel which is the "social implications" of this type of research. I would have thought that the title implied some involvement of non-scientific personnel.

I would also like to comment on Dr. Sinsheimer's observation that the public does not perceive itself as being involved in a dialogue with the scientific community. I feel it is a very true perception on his part.

I am trained as a scientist, but I am not practicing in a laboratory setting at this time; I do sit on a human research committee in a major university; I am involved with an independent, private group of citizens in the same community attempting to ensure adequate health care delivery to all citizens in the county. I am also involved with a bioethics group concerned about the possible institution of this type of research at that university - which is not taking place now. Our particular interest is that non-interested community members, particularly non-scientifically trained members, be involved in that dialogue as to whether this type of research shall be conducted in that community at all, be brought into that dialogue in its initial stages.

It is my observation that in spite of the Cambridge incident, and the activities which occurred more recently in Ann Arbor, the general public wants to be involved in the decisions as to what level of research might take place and in an ongoing dialogue as to whether it should take place at all.

Another point I would like to make is that this discussion has dealt almost entirely with the safety and health aspects of this type of research. There has been no reference whatsoever to the competition which might result due to the limited resources available. Quite frankly, prejudiced as I am as a biologist to this type of activity, I would be quite happy to see funding provided for this type of research on a more wide-spread basis around the United States, but I do not think it is fair to exclude (and I don't think it has been intentional) the general public from those kinds of decision-making processes.

Dr. Singer:
In regard to the point about competition of funds for different kinds of experimentation, I think one of the things about recombinant DNA research is that it is relevant and useful in a very wide range of biological and medical research. As such it is hard to discuss it in the context of competing with other kinds of research because there are so many different fields to which this would be applicable and which would benefit from it.

In regard to the public involvement - I guess it is hardly a question of how much is enough. Clearly for some people, no matter how much public involvement and discussion there is it won't be sufficient; for others, any amount is too much. But, it seems to me that the fact that there were hearings in Congress, which is the representative of the people of this country, and those hearings seemed to conclude that the general process and direction that the issue was taking appeared to be satisfactory, should offer some comfort that there has been a very relevant and responsible public involvement - not to mention all of the discussions in a variety of places throughout the country.

Roy Curtiss, University of Alabama:

Within his talk, Dr. Davis considered many of the probabilities of risks of recombinant DNA research - a topic that I have addressed myself to on numerous occasions. More specifically, what I would like to do is to present a little information in reference to Dr. Davis' comments on the attenuation of laboratory strains such as Escherichia coli K12. As many know, during the past two years our group has been involved in trying to design strains for safer recombinant research and at the same time conducting experiments to evaluate whether some of the biohazards might be real.

While constructing numerous derivative strains during these past two years, we discovered a mutant that has partially restored K12's ability to make a little polysaccharide side chain, more like that made by E. coli strains found in nature that do have the ability to colonize the gut. This mutant does not have a completely smooth phenotype however and has not yet been tested for the ability to colonize animals. The discovery of this mutant, nevertheless, fits in with the long-known phenomenon that laboratory attenuated organisms can often mutate or revert back to virulence. I have further preliminary information to indicate that this is true. For example, we have been feeding strains of Escherichia coli K12 to rodents off and on for over 10 years. We, like others, have not observed persistent colonization of any animal host up until 2 months ago. More recently however we have commenced to feed 10^{10} E. coli cells suspended in milk to rats, thus increasing the dose and blocking the barrier due to stomach acidity.

In an experiment initiated in November of this past year we found three rats that have maintained titers of 10^4 to 10^5 prototropic E. coli K12 cells persistently per gram of feces. This has been for a period now of some eight weeks. We have not yet analyzed these K12 cells from these rats for any alterations in their cell surface that might explain their persistence, nor have we fed these strains to rats or humans not previously exposed to E. coli K12 to see if they might now have the capability of routine colonization and I hasten to add that these observations are extremely preliminary, but I think their ramifications, if borne out by further experiments, raise questions that have already been raised by numerous offers.

Dr. Davis:

Yes, I think this is very interesting information. It suggests that one can really overwhelm the natural flora of the gut by a sufficiently large dose. A very quick calculation suggests to me the 10^{10} organisms in a rat would be about equivalent to a gallon of a full-grown culture in a person.

V. Sgaramella, Laboratorio di Genetica, Pavia:

It seem to me that an effort by the scientific community is in order to find out which key experiments could yield an estimate of these conjectural risks we are talking about. Therefore, I have been circulating a letter to most of the prominant scientists involved in this kind of research asking them to suggest experiments likely to provide answers to this question, whether they could be done in existing facilities, by whom and when, and how long it would take to have results which could be evaluated by the entire scientific community. The letter has been sent to many people and the action which the WHO is considering, following a positive answer (if there is any), would call for a consultant committee which would examine the suggestions and find out how these key experiments could eventually be done. In addition, since figures bearing on the extent of risks are probably not very meaningful unless they are set against some evolutional benefit, I think it would be appropriate to consider experiments directly aimed at giving assurance that practical applications of the genetic recombination technology are taken into proper consideration.

Leon Jacobs, National Institutes of Health:

I am responding to the comment just made and also to a question which was handed up to the panel earlier about what kind of experiments are being done to test the hazards of DNA recombinant molecules.

I would first like to say that the NIAID (National Institute of Allergy and Infectious Diseases) of NIH is also charged with developing a program to examine hazards related to DNA recombinants, and they have put out some Requests for Proposals for work by contract to do some types of testing. In addition to that, there is a committee working to develop a workshop at which it is hoped that appropriate types of experiments will be designed so that we can test a variety of these things. There is a recognition of a need for this, but there are such a variety of things that can be done that it requires a group of people involved in infectious diseases, knowledgeable about such things as infectivity, virulence, pathogenicity, and so on, who can design some cogent experiments to test the various types of recombinants that people will be working with.

Dr. Stetten:

Thank you. I would like to thank the members of the audience - particularly those members who managed to survive throughout this lengthy discussion. I also want to thank the members of the panel. I would like to now return the microphone to our host, Dr. Schultz.

Dr. Schultz:

I would like to acknowledge Dr. Whelan's help in organizing this panel. As you all know, he is the Chairman of the Committee of the International Union of Scientific Societies, who have a committee similar to this one which follows the biohazards, and protection to biohazards, around the world. I would like to express my personal thanks to the panel members and Dr. DeWitt Stetten.

TRANSFECTION OF PERMISSIVE MONKEY CELLS WITH RESTRICTION ENDONUCLEASE-DERIVED FRAGMENTS OF SV40 DNA. I. TRANSFORMATION AND SUPERINFECTION PROPERTIES OF RECIPIENT CELLS.

R.C. Moyer, M.P. Moyer and H. Hurtado,
Trinity University, San Antonio, Texas

Permissive CV1/TC7 cells were separately transfected with SV40 *EcoRI/HapII* A (74% genome) DNA fragments and *EcoRI/HapII* B (26% genome) DNA fragments in the presence of DEAE-dextran. Progeny of recipient cells receiving either the A fragment or the B fragment were capable of growth in soft agar medium whereas mock-inoculated and parental CV1/TC7 cells were not. Tests for infectious SV40 virions in recipient or progeny cultures or their supernatant media by inoculation of permissive cells, by double agar immunodiffusion or by V-antigen immunofluorescence were negative. Cells receiving the A fragments displayed a typical SV40 nuclear T antigen. Cells receiving the B fragments displayed a distinctively cytoplasmic "neo"-antigen which could be detected using SV40 T antiserum. Although only a few percent of either A fragment or B fragment-inoculated recipient cells displayed antigen, sublines derived from soft agar clones (of the A fragment cells) were 80% to 90% T antigen positive. However, only 40% to 60% of the cloned B fragment inoculated cells maintained their positive response to SV40 T antiserum. Uncloned cells containing the A fragment were capable of forming fibrosarcomas when inoculated into neonate hamsters in 5 months. Inoculation of in vitro grown A fragment tumor tissue into neonate hamsters resulted in tumors in 1 month. Cells containing the B fragment were non-tumorigenic. Cloned cells containing either the A fragment or the B fragment were resistant to superinfection by SV40 virions. Superinfection resistance varied between 5% and 100% for cells containing the A fragment and between 48% and 100% for cells containing the B fragment. The data suggest that both the A fragment and the B fragment are capable of transforming permissive cells.

Research sponsored by Thorman Cancer Research Trust

TRANSFECTION OF PERMISSIVE MONKEY CELLS WITH RESTRICTION ENDONUCLEASE-DERIVED FRAGMENTS OF SV40 DNA. II. RESCUE OF SV40 VIRIONS BY CELL FUSION AND TRANSFECTION.

M. Moyer, R. Moyer, M. Gerodetti, and G. Lipotich
Trinity University, San Antonio, Texas

We have previously demonstrated (1,2) that transfection of SV40 74% and 26% subgenomic fragments, generated by restriction endonucleases *Eco*RI and *Hap*II, into permissive African green monkey kidney cells (CV1/TC7) results in stable transformation and superinfection resistance. Infectious SV40 virions could be rescued when cells containing the 26% genome fragment were fused, using inactivated Sendai virus, with cells containing the 74% genome fragment. Virus could be rescued when DNAs from cells containing the 74% genome fragment and cells containing the 26% genome fragment were transfected into CV1/TC7 cells. No other combination of fragment-transformed cells and/or permissive cells yielded infectious virions following cell fusion or transfection. Cocultivations of cloned cells containing the 74% genome fragment with cloned cells containing the 26% genome fragment did not yield infectious SV40 virions. Viral rescue was evidenced by SV40 cytopathology and V antigen production. Plaque assay, neutralization tests, and electron microscopy further verified that the rescued virus was SV40. These studies suggest: (1) the subgenomic fragments are present only in their respective cell lines; (2) the fragments are transmittable to progeny cells through numerous subcultures; and, (3) the fragments are not only present in an intact form but remain potentially biologically active.

REFERENCES

(1) R.C. Moyer and M.P. Moyer, 1976. Trans. Gulf Coast Mol. Biol. Conf. TJS Spec Publ 1:59-72.
(2) R.C. Moyer, M.P. Moyer, and M.H. Gerodetti, 3rd International Symposium on Detection and Prevention of Cancer, New York (1976).

IN VITRO RECOMBINANT DNA TECHNOLOGY FOR MAPPING ANIMAL VIRUS MUTATIONS

Lois K. Miller, Department of Bacteriology and Biochemistry, University of Idaho, Moscow, Idaho 83843 and the Imperial Cancer Research Fund, London, England

The formation of hybrid viruses by *in vitro* recombination of specific restriction endonuclease (REN) fragments of polyoma DNA followed by infection of secondary mouse embryo cells with the recombinant DNA has proved to be a useful technique for localizing genetic mutations on the polyoma DNA genome. The formation of two hybrid viruses, each containing approximately one-half of the genetic information of two different strains of polyoma virus, provided a prototype for genetic mapping by recombinant DNA technology (1). Since the two parental polyoma strains differed phenotypically (plaque morphology and HA properties) and genotypically (REN fragment pattern variations), the formation of hybrid viruses was demonstrated by REN analysis and the assignment of the DNA fragment controlling plaque morphology and HA properties was made by analyzing the phenotypes of the two hybrid viruses. This general technique has been applied to the mapping of temperature sensitive mutants of polyoma, including a late mutant ts59. By cleaving both ts59 and a wild type (wt) virus into two fragments with the REN $Hind_{III}$, purifying the fragments by gel electrophoresis, recombining appropriate fragments *in vitro*, ligasing the fragments together and infecting mouse cells with the hybrid DNAs, two hybrid viruses were constructed. One hybrid virus was temperature sensitive for infection at the non-permissive temperature and was composed of the larger $Hind_{III}$ DNA fragment of ts59 and the smaller $Hind_{III}$ DNA fragment of the wt virus. Further biochemical genetic studies and peptide analysis of VP1 and VP2 (capsid proteins) of the parental and hybrid viruses have provided specific information concerning the location of the VP1 and VP2 in the polyoma genome. Such *in vitro* recombinant DNA techniques permit genetic manipulation of animal virus DNA genomes, mapping of viral mutations and correlation of viral proteins with regions of the virus genome.

REFERENCES

(1) L. K. Miller and M. Fried (1976) Nature 259:598-601 "Construction of Infectious Polyoma Hybrid Genomes *in vitro*.

NUCLEAR TRANSLATIONAL UNITS OF ADENOVIRUS-INFECTED HeLa CELLS

N.K. Chatterjee, H.W. Dickerman and T.A. Beach,
Division of Laboratories and Research, New York State Department of Health, Albany, New York.

A separate and discrete population of polyribosomes exists in the detergent-washed nuclei of adenovirus-infected HeLa cells. These polyribosomes, released by exposure to polynucleotides such as hnRNA or poly(U), do not appear to be cytoplasmic contaminants. Nuclear polyribosomes have a considerably lower buoyant density compared to cytoplasmic ones. Nuclear polyribosomes, in a cell-free system of protein synthesis, are 6- or 8-fold less active compared to cytoplasmic ones and are insensitive to aurin tricarboxylic acid. They do not complement cytoplasmic polyribosomes in protein synthesis in the cell-free system. Finally, the number of proteins synthesized by nuclear polyribosomes is higher compared with that synthesized by the cytoplasmic ones. Only the virus specific proteins, including P-VII, are synthesized by cytoplasmic polyribosomes. Nuclear polyribosomes, on the other hand, synthesize virus specific proteins, including P-VII and VII, and a number of additional proteins not synthesized by the cytoplasmic ones. Experimental evidence tends to support the suggestion that nuclear polyribosomes may take part in the synthesis or processing of adenovirus proteins.

Supported in part by GRS Grant 5S01RR-05649-09 and by Research Grant 1R01-CA19707-01 from the NCI, DHEW.

TRANSCRIPTION AND COUPLED TRANSCRIPTION-TRANSLATION OF CLONED DNAs IN XENOPUS OOCYTES.

J.E. Mertz, J.B. Gurdon, and E.M. De Robertis,
McArdle Laboratory for Cancer Research, Madison, Wis., and
MRC Laboratory of Molecular Biology, Cambridge, England.

The African clawed frog, *Xenopus laevis*, has been widely employed in a variety of studies concerned with control of eucaryotic gene expression. With the findings that: 1) microinjected cultured cell nuclei are transcriptionally active for weeks (Gurdon *et al.*, *Nature*, *260*: 116 (1976); 2) purified SV40 DNA becomes reconstituted into chromatin-like structures following microinjection into eggs (R. Laskey, A. Mills, and N.R. Morris, submitted); 3) transcription of ribosomal and 5S gene DNA can be detected after injection into fertilized eggs (J.B. Gurdon and D.D. Brown, in press); and 4) injected mRNAs are translated in oocytes, we have been encouraged to look at transcription and coupled transcription-translation of purified DNAs in oocytes and eggs.

When SV40 DNA is microinjected into the nucleus of *Xenopus* oocytes subsequently incubated with ^{3}H-uridine or injected with ^{3}H-GTP, radioactivity is incorporated into DNase-resistant, RNase- and alkali-sensitive material that specifically hybridizes to SV40 DNA. This viral transcription: 1) is the predominant species of labeled non-ribosomal RNA made when 5 ng or more of SV40 DNA is injected per oocyte; 2) continues unabated for days; 3) only occurs when the DNA is injected *directly* into the nucleus of the oocyte; and 4) occurs, although less efficiently, following injection of the DNA into *Xenopus* eggs. Using long labeling periods, we find that some of the virus-specific RNA produced in oocytes resembles the viral mRNAs made during lytic infection of monkey cells. In addition, some new proteins, possibly coded by the virus, are also made in oocytes injected with SV40 DNA. These results suggest that the oocytes may be processing the viral transcripts into functional mRNAs. The DNAs of Adenovirus 5, cloned *Drosophila* histone genes, and even ØX174 RF, Ø80p*lac* and ColE1 are also transcribed in this system. We are presently testing whether this system is useful: 1) as an *in vivo* transcription and coupled transcription-translation system for identifying and mapping the mRNAs and proteins made from cloned DNAs; 2) for defining and localizing promotors for transcription of eucaryotic genes; and 3) as an assay system for the functional identification of "factors" involved in the control of initiation of transcription.

A SAFER MODEL SYSTEM FOR STUDYING THE EFFECTS OF RECOMBINING ANIMAL VIRUS DNA

Lois K. Miller, Department of Bacteriology and Biochemistry, University of Idaho, Moscow, Idaho 83843

A number of researchers are interested in determining the potential usefulness of recombinant DNA technology to explore a variety of aspects of animal viruses including their genetic organization, host interactions and virulence. Such recombinant DNA experimentation presents a number of potential hazards, particularly the recombination of DNA from viruses known to infect vertebrates. There are several animal viruses currently available which are safer model systems for such recombinant studies. Certain insect baculoviruses have recently been registered by the Environmental Protection Agency for use as biological pesticides. Registration of these viruses for pesticide purposes (massive dissemination by spraying) has required extensive testing of these viruses for safety with regard to a wide range of animal and plant species. A number of invertebrate, avian, piscian and mammalian (including two primate) species have been subjected to baculovirus administration by a variety of routes including inhalation, eye application, skin abrasion, per os and injection (subcutaneous, intradermal, intramuscular, intracerebral, intravenous and intraperitoneal). Tests for teratogencity, carcinogenicity and replication in tissue culture have been performed with various infectious forms of the virus including infectious DNA. Humans have been fed large quantities of baculovirus with no observable effect - in fact, humans normally consume large quantities of baculoviruses as contaminants of cole crops. The viruses are species or genus specific and their hosts are insect pests. If certain potential hazards of genetic engineering prove to be true, the virtual elimination of the pest species would merely eliminate a pest problem. Although no virus can be deemed unequivocally safe (as is also true for any given genetic engineering experiment), the relative safety features of baculoviruses approved by the Environmental Protection Agency for pesticide use indicate these viruses are preferable model systems for determining the usefulness of recombinant DNA technology in virus studies. The properties of these viruses, their safety and their potential use for genetic engineering studies will be discussed.

A ROLE FOR CELL SURFACE TOPOGRAPHY FOR DIRECTING THE SYNTHESIS OF ANTI-GLYCOSYL ANTIBODIES

John H. Pazur and Kevin L. Dreher,
Department of Biochemistry and Biophysics, Pennsylvania State University, University Park, Pa., 16802

Recently, two types of anti-glycosyl antibodies which are specific for different structural moieties of the same carbohydrate antigen have been isolated by affinity chromatography. The carbohydrate antigen is a Group D streptococcal diheteroglycan consisting of a main chain of glucose-glucose-galactose repeating units with lactose side chains linked to the second glucose unit. The anti-glycosyl antibodies were isolated from antisera of rabbits immunized with a vaccine of non-viable Streptococcus faecalis cells with the antigen in situ in the cell wall matrix. Affinity chromatography was performed on lactosyl-sepharose columns with galactose and lactose solutions as the eluting agents. The antibodies which were eluted with galactose are anti-galactose antibodies while those eluted with lactose are anti-lactose antibodies. The anti-galactose preparation consists of 6 distinct proteins, while the anti-lactose preparation consists of 10 different proteins. Since the proteins of each set combine with the same structural unit of the antigen they are appropriately termed isoantibodies. The isoantibodies are separable into individual components by gel-electrophoretic and electrofocusing techniques. Rabbits immunized with galactosyl-BSA produced only the anti-galactose antibodies while rabbits immunized with lactosyl-BSA produced only anti-lactose antibodies. The latter result was surprising since the immunocytes of the animal were in contact with the same carbohydrate side chains in both the synthetic and the natural antigen. These findings on the synthesis of isoantibodies with the natural and synthetic antigens have been interpreted to indicate a role for the topography of the bacterial surface in the synthesis of antibodies. Because of a particular arrangemtnt of the immunodeterminant groups on the cell surface two types of anti-glycosyl antibodies were synthesized by the immunocytes in contact with the natural antigen. However since galactosyl-BSA or lactosyl-BSA can not exhibit the desired topography only one set of anti-glycosyl antibodies is produced with these immunogens.

A
B 7
C 8
D 9
E 0
F 1
G 2
H 3
I 4
J 5